Food Analysis in the 21st Century: Challenges and Possibilities

Food Analysis in the 21st Century: Challenges and Possibilities

Editor

Gavino Sanna

MDPI

Basel • Beijing • Wuhan • Barcelona • Belgrade • Novi Sad • Cluj • Manchester

Editor
Gavino Sanna
Department of Chemical,
Physical, Mathematical and
Natural Sciences
Sassari University
Sassari
Italy

Editorial Office
MDPI
St. Alban-Anlage 66
4052 Basel, Switzerland

This is a reprint of articles from the Special Issue published online in the open access journal *Molecules* (ISSN 1420-3049) (available at: www.mdpi.com/journal/molecules/special_issues/Food_Analysis_21st).

For citation purposes, cite each article independently as indicated on the article page online and as indicated below:

Lastname, A.A.; Lastname, B.B. Article Title. *Journal Name* **Year**, *Volume Number*, Page Range.

ISBN 978-3-7258-0768-0 (Hbk)
ISBN 978-3-7258-0767-3 (PDF)
doi.org/10.3390/books978-3-7258-0767-3

Contents

About the Editor vii

Preface ix

Colleen L. Ray, James A. Gawenis and C. Michael Greenlief
A New Method for Olive Oil Screening Using Multivariate Analysis of Proton NMR Spectra
Reprinted from: *Molecules* **2022**, *27*, 213, doi:10.3390/molecules27010213 1

Agnese Giacomino, Paolo Inaudi, Gessica Silletta, Aleandro Diana, Stefano Bertinetti, Elisa Gaggero, et al.
Analytical Methods for the Characterization of Vegetable Oils
Reprinted from: *Molecules* **2023**, *28*, 153, doi:10.3390/molecules28010153 14

Evangelia T. Ioannou, Konstantinos S. Gliatis, Evangelos Zoidis and Constantinos A. Georgiou
Olive Oil Benefits from Sesame Oil Blending While Extra Virgin Olive Oil Resists Oxidation during Deep Frying
Reprinted from: *Molecules* **2023**, *28*, 4290, doi:10.3390/molecules28114290 31

Fugang Xiao, Menglin Gu, Yaoxuan Zhang, Yaodong Xian, Yaotian Zheng, Yongqing Zhang, et al.
Detection of Soybean-Derived Components in Dairy Products Using Proofreading Enzyme-Mediated Probe Cleavage Coupled with Ladder-Shape Melting Temperature Isothermal Amplification (Proofman–LMTIA)
Reprinted from: *Molecules* **2023**, *28*, 1685, doi:10.3390/molecules28041685 42

Andrea Mara, Marco Caredda, Margherita Addis, Francesco Sanna, Mario Deroma, Constantinos A. Georgiou, et al.
Elemental Fingerprinting of Pecorino Romano and Pecorino Sardo PDO: Characterization, Authentication and Nutritional Value
Reprinted from: *Molecules* **2024**, *29*, 869, doi:10.3390/molecules29040869 54

Siwei Wang, Xiaonan Wang, Qiang He, Haidan Lin, Hong Chang, Haibin Sun and Yanping Liu
Simultaneous Determination of Seven Pesticides and Metabolite Residues in Litchi and Longan through High-Performance Liquid Chromatography-Tandem Mass Spectrometry with Modified QuEChERS
Reprinted from: *Molecules* **2022**, *27*, 5737, doi:10.3390/molecules27175737 69

Michele Protti, Isacco Gualandi, Sergio Zappoli, Roberto Mandrioli, Laura Mercolini and Domenica Tonelli
Evaluation of the Antioxidant Capacity of Fruit Juices by Two Original Analytical Methods
Reprinted from: *Molecules* **2023**, *28*, 6672, doi:10.3390/molecules28186672 80

Sergio Carneado, José Fermín López-Sánchez and Ángeles Sahuquillo
Antimony in Polyethylene Terephthalate-Bottled Beverages: The Migration Puzzle
Reprinted from: *Molecules* **2023**, *28*, 7166, doi:10.3390/molecules28207166 88

João Victor Basolli Borsatto, Edvaldo Vasconcelos Soares Maciel, Alejandro Cifuentes and Fernando Mauro Lanças
Applicability and Limitations of a Capillary-LC Column-Switching System Using Hybrid Graphene-Based Stationary Phases
Reprinted from: *Molecules* **2023**, *28*, 4999, doi:10.3390/molecules28134999 104

Esmanur İlhan, Pelin Poçan, Danuta Kruk, Miłosz Wojciechowski, Maciej Osuch, Roksana Markiewicz, et al.
Water Dynamics in Starch Based Confectionery Products including Different Types of Sugar
Reprinted from: *Molecules* **2022**, *27*, 2216, doi:10.3390/molecules27072216 . . . . **118**

Luke Taylor, Lili Saskői, Tara Brodie, Vytautas Remeškevičius, Hannah Jayne Moir, James Barker, et al.
Development of a Gas-Tight Syringe Headspace GC-FID Method for the Detection of Ethanol, and a Description of the Legal and Practical Framework for Its Analysis, in Samples of English and Welsh Motorists' Blood and Urine
Reprinted from: *Molecules* **2022**, *27*, 4771, doi:10.3390/molecules27154771 . . . . **128**

About the Editor

Gavino Sanna

Gavino Sanna is currently an Associate Professor of Analytical Chemistry at the University of Sassari, Sardinia, Italy. After five years at EniChem Research Center, Porto Torres, he became Assistant Professor. In 2018, he obtained the Abilitazione Scientifica Nazionale as Full Professor in Analytical Chemistry. With more than 100 publications in international journals, books, and patents, his research focuses on instrumental methods for the analysis of organic and inorganic analytes in food, materials, the environment, and biological samples. Major contributions include the determination and speciation of minority analytes in traditional Sardinian foods, the influence of irrigation methods on the bioaccumulation of health-threatening elements in rice, and electroanalytical studies of metal ion–biological ligand interactions. His expertise extends to food safety, environmental analysis, materials science, and biomedical analysis. He is the Associate Editor of MDPI's *Separations* journal and Elsevier's *Journal of Agriculture and Food Research*, as well as Guest Editor of several Special Issues and Topics published by MDPI (*Molecules, Separations, Foods, Processes,* and *Analytica*) and Elsevier (*Journal of Agriculture and Food Research*). He is also a reviewer for several dozen renowned Q1 international journals in the fields of chemistry, analytical chemistry, environmental sciences, food science and technology, as well as food analysis.

Preface

This reprint of the Special Issue "Food Analysis in the 21st Century: Challenges and Possibilities" presents eleven regular papers. The papers cover current topics in food analysis in the 21st century. The methods proposed are based on analytical techniques such as inductively coupled plasma or chromatographic approaches, used alone or combined with mass spectrometry, nuclear magnetic resonance, electroanalytical techniques, and DNA-based methods. They have been applied to matrices as diverse as edible oils, dairy products, water, beverages, fruits, and saccharide-based foods for the determination of toxic substances, trace elements, and nutraceutical species. The aims include food characterization, food safety protection, food authentication or adulteration, and the determination of recreational substances of abuse. The use of advanced chemometric techniques has often been crucial in maximizing analytical information to achieve results in the classification and detection of species that evidence the adulteration or pollution of food productions. This volume wishes to be a valuable resource for all analytical chemists, particularly those interested in embarking on the challenging yet captivating journies in analytical food chemistry.

Gavino Sanna
Editor

molecules

MDPI

Article

A New Method for Olive Oil Screening Using Multivariate Analysis of Proton NMR Spectra

Colleen L. Ray [1], James A. Gawenis [2] and C. Michael Greenlief [1,*]

[1] Department of Chemistry, University of Missouri, 601 S. College Avenue, Columbia, MO 65211, USA; clrxtc@mail.missouri.edu
[2] Sweetwater Science Laboratories, Glasgow, MO 65264, USA; jim@sweetwaterscience.com
[*] Correspondence: greenliefm@missouri.edu; Tel.: +1-573-882-3288

Abstract: A new NMR-based method for the discrimination of olive oils of any grade from seed oils and mixtures thereof was developed with the aim of allowing the verification of olive oil authenticity. Ten seed oils and seven monovarietal and blended extra virgin olive oils were utilized to develop a principal component analysis (PCA) based analysis of ^{1}H NMR spectra to rapidly and accurately determine the authenticity of olive oils. Another twenty-eight olive oils were utilized to test the principal component analysis (PCA) based analysis. Detection of seed oil adulteration levels as low as 5% v/v has been shown using simple one-dimensional proton spectra obtained using a 400 MHz NMR spectrometer equipped with a room temperature inverse probe. The combination of simple sample preparation, rapid sample analysis, novel processing parameters, and easily interpreted results, makes this method an easily accessible tool for olive oil fraud detection by substitution or dilution compared to other methods already published.

Keywords: proton NMR; food authenticity; adulteration; olive oil; edible oil; PCA

Citation: Ray, C.L.; Gawenis, J.A.; Greenlief, C.M. A New Method for Olive Oil Screening Using Multivariate Analysis of Proton NMR Spectra. *Molecules* **2022**, *27*, 213. https://doi.org/10.3390/molecules27010213

Academic Editor: Gavino Sanna

Received: 13 December 2021
Accepted: 29 December 2021
Published: 30 December 2021

1. Introduction

Olive oil is the oil collected from the fruit of the olive tree (*Olea europaea* L.) typically through simple mechanical pressing. Olive oil is somewhat unusual as the oil is extracted from the flesh of the fruit instead of the seed as is common in most other food oils. This oil has been consumed by humans since antiquity and remains a highly valued food oil today. Due to the high market price of olive oil compared to other oils, it is a popular target for adulteration through dilution with other oils or label fraud by selling non-olive sourced oils as genuine olive oil [1]. The aim of this work is to develop a rapid analysis for the detection of seed oil adulteration in any grade of olive oil.

The motivation for adulteration is one of simple greed. If olive oil is diluted with a less costly oil or is completely replaced by said oil, the profits from selling it as genuine olive oil can be quite large. Adulteration of olive oils can affect consumers beyond the obvious economic impact of paying a premium for fraudulent goods. Olive oil is often consumed for its reputed health benefits due to its unique composition, which would be reduced if diluted or absent if the product contains no olive oil whatsoever. Inadvertent consumption of oils ordinarily avoided by people with allergies could have significantly more serious and immediate effects on the consumer if the so-labeled olive oil contains products to which the consumer is sensitive.

NMR spectroscopy has long been the gold standard method for the elucidation of unknown molecular structures and is often used in synthetic chemistry for verification of products. In recent years the ability for NMR spectroscopy to screen products and materials for quality control or authenticity has gained significant attention. The data analysis methods employed in this study are broadly similar to those used for untargeted metabolomic fingerprinting commonly performed with mass spectrometry. While mass spectrometry has greater sensitivity, NMR is capable of more rapidly screening complex mixtures such as foodstuffs in a non-destructive manner. Coupling NMR results with principal component analysis (PCA) allows even subtle differences in overall composition to be useful for discriminating between oil sources and detecting adulterated samples. Utilizing a 400 MHz NMR to detect lower levels of adulteration of olive oil with oils such as high-oleic acid cultivars of sunflower and safflower oil is challenging due to these adulterants having lipid profiles very similar to those found in olive oils [2].

Olive oil is composed of fatty acid triglycerides with lower concentrations of a variety of phenolic and polyphenolic compounds [3]. Oleic acid is the most abundant fatty acid found in olive oils with varying levels of linoleic acid, linolenic acid, and palmitic acid. The ratios of the various fatty acids in olive oil differs from those found in many seed oils, particularly due to the high levels of oleic acid and low levels of ω-3 α-linolenic acid. The spectral signature of the lower concentration fatty acids contributes significantly to differentiating olive oils from high oleic acid varietals of seed oils.

These differing levels manifest themselves spectrally and allow these oils to be differentiated via analysis. A spectral comparison of hempseed oil and a sample of monovarietal picual olive oil demonstrates many of these differences [4] (Figure 1). The triplet "F" at 0.977 ppm arises from the terminal methyl group of ω-3 fatty acids, most often α-linolenic acid, which is found in significantly higher concentrations in hempseed oil than olive oil. The proximity of the π-bond between carbons 2 and 3 in ω-3 fatty acids deshields the terminal methyl group resulting in a shift away from peak "G" at 0.881 ppm belonging to the same functional groups in other fatty acids.

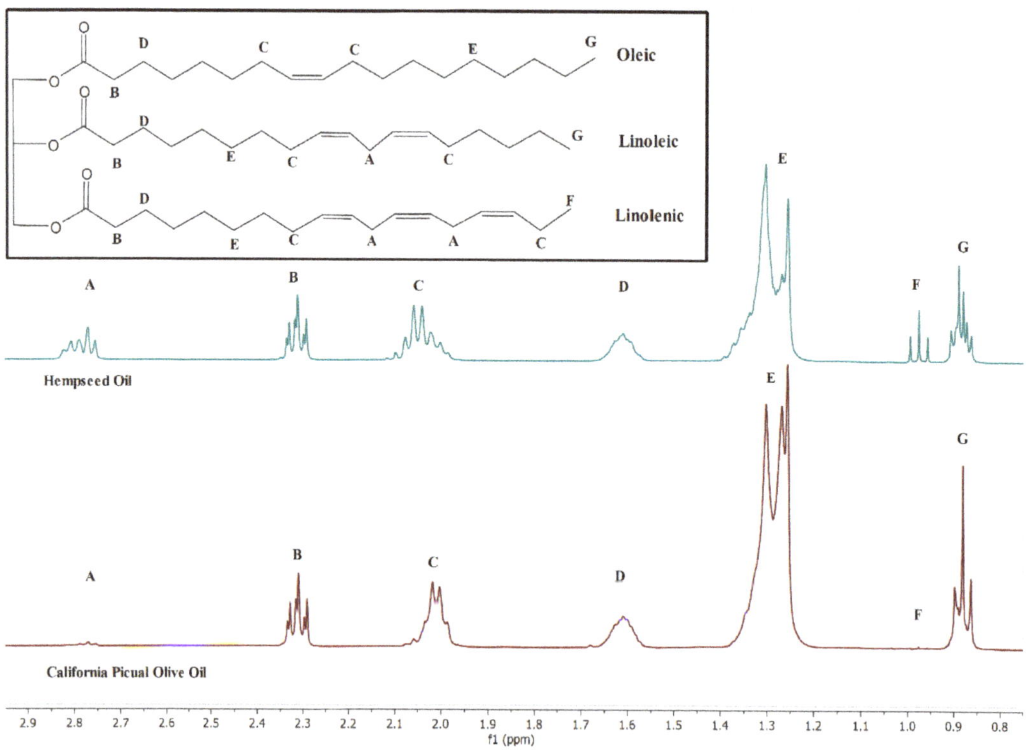

Figure 1. Comparison of 400 MHz ^{1}H NMR spectra of hempseed and olive oil with model triglyceride structure detailing origins of specific labeled peaks from various fatty acid subgroups.

Multiplet "A" at 2.78 ppm (Figure 1) is due to the methylene protons positioned between two π-bonds in polyunsaturated fatty acids, such as linoleic and linolenic. Polyunsaturated fatty acids are plentiful in hempseed oil and are in relatively low abundance in olive oils as seen by the intensity difference of this feature in the two spectra [4].

Analyzing these spectra manually via peak area ratiometrics is possible but is time intensive and tedious due to the number of peaks and variables involved. Utilizing PCA to group similar spectra together accomplishes a similar overall goal while being far easier to automate and produces easily interpreted results [5]. PCA is often used in many complex analyses and has been shown here to work well to discriminate between spectra of various food oils. Not only is the method described herein able to differentiate pure oils by source, but the method can also detect olive oil that has been diluted by other oils.

Olive oil authenticity testing via gas chromatography (GC) and liquid chromatography (LC) are well established methods. These GC and LC methods are comparatively slow, often involving several sample preparation steps, and sample analysis runs on the order of 30 min with longer run times being commonplace [6,7]. NMR analysis of oil samples requires no sample preparation aside from mixing the sample with a deuterated solvent and experiment run times are on the order of 15 min for a simple one-dimensional proton NMR. PCA of food oil NMR spectra is not entirely a novel development in and of itself [8,9]. However, previous studies did not demonstrate the ability to detect adulteration via dilution, and generally dealt with differentiating olive oils by geographical location. These studies also used NMR spectrometers with higher field magnets that are more expensive and less widely available than the comparatively inexpensive 400 MHz system used in this study.

2. Materials and Methods

2.1. Chemicals and Materials

Deuterated chloroform ($CDCl_3$ 99.9% D, 1% *w*/*w* TMS) was obtained from Acros Organics (Fair Lawn, NJ, USA).

2.2. Oil Samples

Olive oil and seed oil samples were purchased from local and online retailers. Olive oil samples consisted of monovarietal and blended oils of European, Mediterranean, and North American origin. A single premixed blend of 70% canola, 20% olive oil, and 10% (all *v*/*v*) grapeseed oil was used as a blended sample for comparison. Table 1 lists the olive oil samples used in this study.

The seed oils used for comparison and adulteration studies were: almond, argan, high-oleic canola, cottonseed, grapeseed, hazelnut, hempseed, peanut, high-oleic safflower, soybean, and high-oleic sunflower oils. All seed oils were purchased from online and local retailers.

2.3. Sample Preparation

50 μL of oil was added directly to a clean 5 mm NMR tube (Deutero Boroeco 8, Deutero GmbH, Kastellaun, Germany) with a pipette. 550 μL of $CDCl_3$ was then added to the NMR tube. The tube was then capped, inverted to ensure complete mixing, and then analyzed.

Adulterated Samples

Sample 20, a coratina monovarietal olive oil from California, was mixed with varying levels of canola, hazelnut, peanut, safflower, and sunflower oils to test the ability of this method to detect adulteration via dilution. Canola and hazelnut oil adulteration samples were prepared with 10%, 15%, and 20% (*v*/*v*) of adulterant. The same olive oil was also adulterated with 10%, 20%, 30%, and 40% peanut, safflower, and sunflower oils. A premixed commercially available blend of 70% canola, 20% olive oil, and 10% grapeseed oil (all *v*/*v*) was also analyzed to further test the model.

Table 1. List of olive oil samples used.

Sample	Varietal	Grade
1	Arbequina	Extra Virgin
2	Picual	Extra Virgin
3	Nocellara	Extra Virgin
4	Manzanillo	Extra Virgin
5	Hojiblanca	Extra Virgin
6	Coratina	Extra Virgin
7	Koroneiki	Extra Virgin
8	Blend	Extra Virgin
9	Manzanillo	Extra Virgin
10	Hojiblanca	Extra Virgin
11	Blend	Extra Virgin
12	Kilkai	Extra Virgin
13	Manzanillo	Extra Virgin
14	Ascolano	Extra Virgin
15	Arbequina	Extra Virgin
16	None Specified	Extra Virgin
17	None Specified	Extra Virgin
18	None Specified	Extra Virgin
19	Pendolino	Extra Virgin
20	Coratina	Extra Virgin
21	Picual	Extra Virgin
22	Coratina	Extra Virgin
23	None Specified	Olive Oil
24	None Specified	Olive oil
25	None Specified	Refined
26	None Specified	Extra Virgin
27	None Specified	Extra Virgin
28	Blend	Extra Virgin

All adulterated samples were prepared to a final volume of 5 mL. Olive oil and adulterants were measured into a 15 mL conical tube, vortexed for 20 s to ensure complete mixing, and prepared for analysis as described in Section 2.3.

2.4. NMR Analysis

NMR spectra were collected using a Bruker Avance IIIHD spectrometer operating at 400.13 MHz. The probe used was a 5 mm BBI room temperature probe. The sample temperature was 298 K. A simple proton experiment was performed (30° pulse, 64 scans, 2 dummy scans, 20 ppm sweep width, 65,536 data points). A 10 s relaxation delay was used in order to ensure complete relaxation between scans based upon a 1 s T_1 measurement.

2.5. Spectral Processing Parameters

Spectral processing was performed with Mestrenova 14.1 (Mestrelab, Santiago de Compostela, Spain). Spectral processing parameters described in Table 2.

Table 2. Spectral processing parameters.

Parameter	Value
Spectral Reference	TMS
Apodization	0.3 Hz
Phase Adjustment	Automatic
Baseline Correction	Polynomial, order = 5
Intensity Normalization	Peak at 0.975 ppm

2.6. Principal Component Analysis

PCA analysis was performed using Mestrenova 14.2 (Mestrelab, Santiago de Compostela, Spain). The PCA analysis was blinded to six regions to eliminate portions of the spectrum irrelevant to oil analysis using the parameters outlined in Table 3. The PCA settings were as follows: Binning mode regular with summed intensity, bin width: 0.05 ppm. Pareto scaling was applied.

Table 3. List of blinded regions used in the PCA analysis.

High Frequency Limit (ppm)	Low Frequency Limit (ppm)	Item Eliminated
−3.9	−1.0	Low frequency noise region
−0.2266	0.1892	TMS peak and satellites
6.995	7.006	$CDCl_3$ satellite peak
7.195	7.325	$CDCl_3$ main peak
7.502	7.548	$CDCl_3$ satellite peak
13.02	16.17	High frequency noise region

3. Results

3.1. Normalization of Spectra

As shown in Figure 2, reasonable grouping of olive oils (green) was observed with principal components 1 and 2 accounting for 97.7% and 1.8% of total variance, respectively. The blue- and magenta-colored ellipses contain information about the adulterated olive oil samples. The percentages shown on the figure indicate the amount of adulterant oil added to olive oil. In the PC1-PC2 plot on the left-hand side, the intensities are normalized to the tallest peak in the spectrum. It is not possible to differentiate these samples easily even at adulteration levels of 40% *v*/*v*. However, when the spectra are normalized to the ω-3 methyl signal, as shown in right hand side of Figure 2, this technique becomes significantly more sensitive to adulteration with these oils and also shows far tighter grouping of olive oil samples. Based on these results, all spectra were normalized to the ω-3 methyl signal in the remainder of the study.

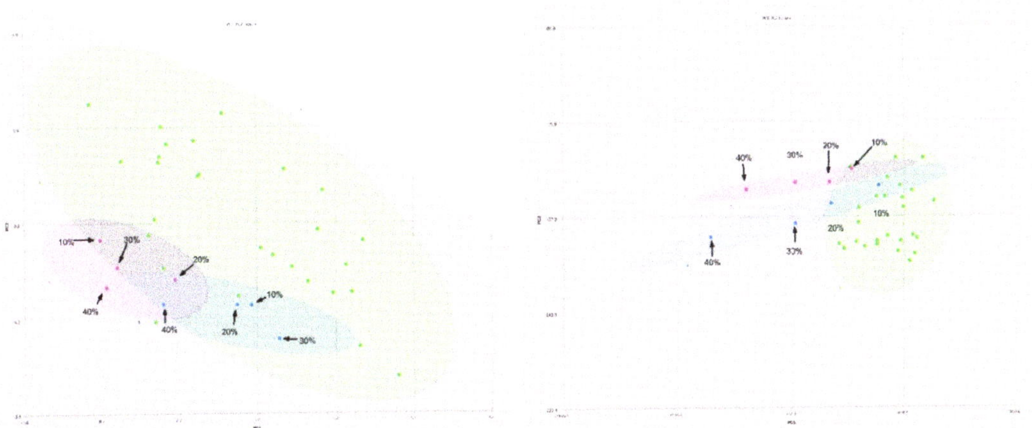

Figure 2. Comparison of PC1–PC2 plots of 400 MHz 1H NMR spectra of pure olive oils (green), safflower oil adulterated olive oil (magenta) and sunflower oil adulterated olive oil (blue) showing the differences between traditional normalization (**left**) and normalization to ω-3 fatty acids (**right**).

3.2. *Differentiation of Olive and Seed Oils*

A total of 28 single varietal and blended olive oils and 10 seed oils were analyzed via NMR with PCA performed on the collected spectra. Figure 3 summarizes the results. The cluster of green dots are the different olive oils samples and are observed in a tight cluster. The seed oils are shown as maroon dots and are clearly separated from the olive oils. There is a cluster of seed oils (maroon) in Figure 3 to the right of the olive oils. This region is expanded in Figure 4. The expanded region clearly shows the differences between hazelnut oil and hempseed oil, for example.

Figure 3. PC1–PC2 plot of 400 MHz ^{1}H NMR spectra of seed oils (labeled) versus olive oils (green). Ellipses represent the 95% confidence interval. The congested region is expanded in Figure 4.

3.3. *Testing against Mixtures of Olive Oil and Seed Oils*

In Figure 5, the green dots represent the same olive oils shown in Figure 2. The extended ellipses represent olive oils samples that have been adulterated as described in the previous section. To the far right of the figure, the labeled maroon dots indicate the location of pure hazelnut oil (also indicated by an arrow in the figure). As the concentration of olive oil is increased the red dots show how the samples moved towards the pure olive oil region (green). Four other oils were used to dilute olive oil. In each case, the undiluted sample (pure olive oil) resides in the green region. As the concentration of the adulterant oil increases, the ellipses move towards the respective oil used to dilute the olive oil (peanut, safflower, sunflower, or canola oil). Due to the greater degree of separation, particularly in the case of high-oleic canola, safflower, and sunflower oils, this allows for identification of the adulterant oil.

Figure 6 shows the results for different concentrations of canola oil compared to different olive oils. Pure canola oil shows excellent separation from pure olive oils as shown in Figure 6 with an ability to discriminate pure olive oils from those adulterated with canola at levels under 10% v/v.

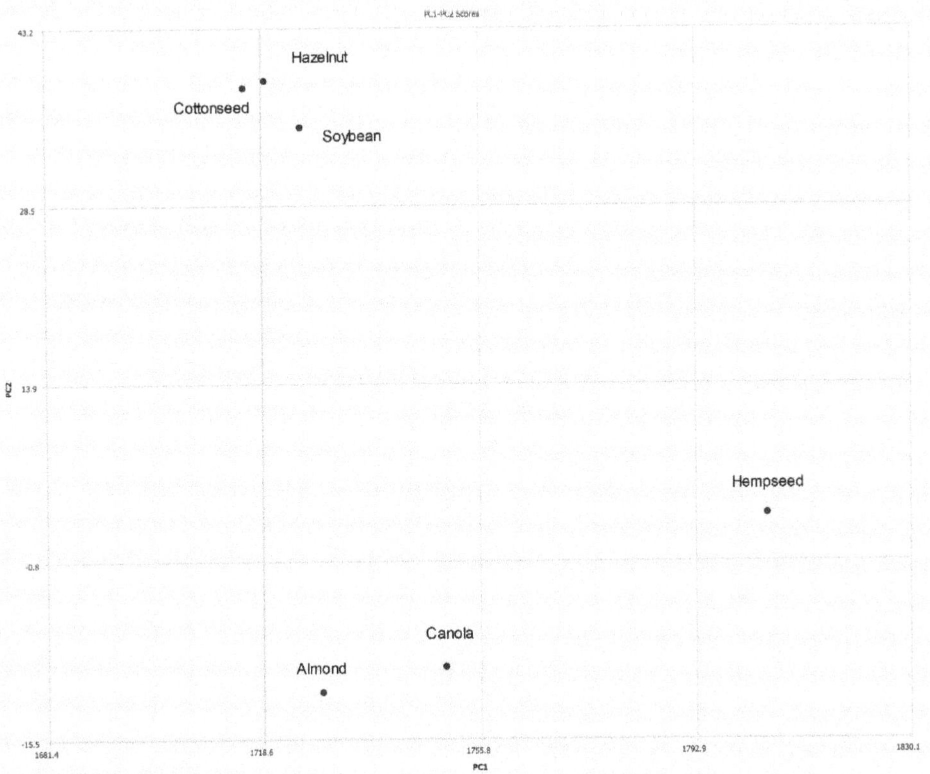

Figure 4. PC1–PC2 plot expansion of congested region in Figure 3.

The results from the same type of experiment, but using hazelnut oil adulterated olive oil is easily detected even at levels as low as 5% v/v are shown in Figure 7. The pure hazelnut oil is easily distinguished from pure olive oil. The adulterated samples following a grouping between the two pure oils as the concentration of adulterated hazelnut oil is varied.

Peanut oil adulterated olive oil is detectable at concentrations below 20% v/v as shown in Figure 8. Again, pure peanut oil is easily separated from pure olive oil samples using the PCA analysis.

High oleic acid safflower oil is ordinarily rather difficult to distinguish from olive oils due to having similar lipid profiles. However, using this method it is detectable in levels slightly under 20% v/v as observed in Figure 9. Samples above this level are readily detected as not being olive oil.

High oleic acid sunflower oil is another ordinarily difficult to detect adulterant of olive oils, yet it is detectable at levels just over 20% v/v using this method as seen in Figure 10. As the concentration of sunflower oil increases, the sample points on the PCA plot trend toward the pure sunflower oil sample. Differences in the adulterated oil will affect the end result as not all olive oil samples fall exactly together on the PCA plot, but this tool will still identify high-oleic sunflower oil adulterated olive oils at economically viable levels.

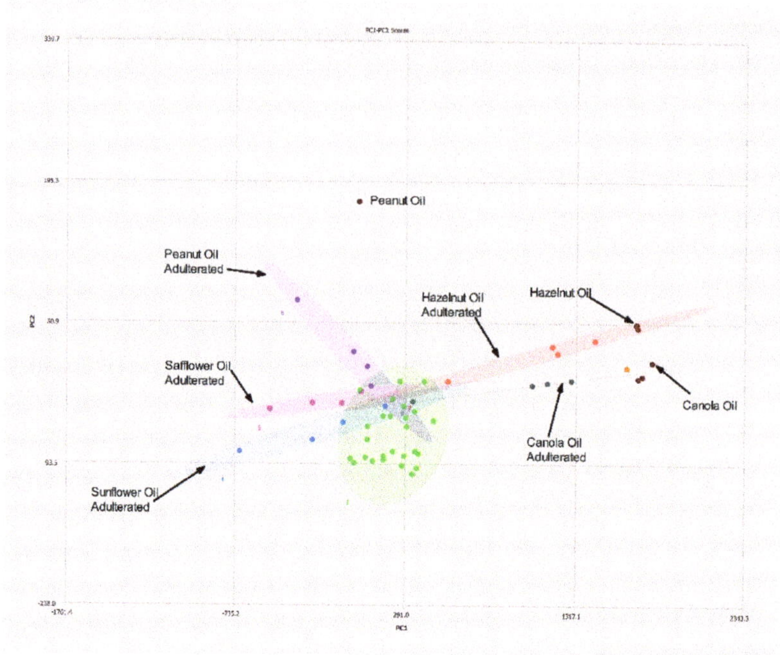

Figure 5. PC1–PC2 plot of 400 MHz ^{1}H NMR spectra of seed oils (labeled) versus olive oils (green). Ellipses represent the 95% confidence interval.

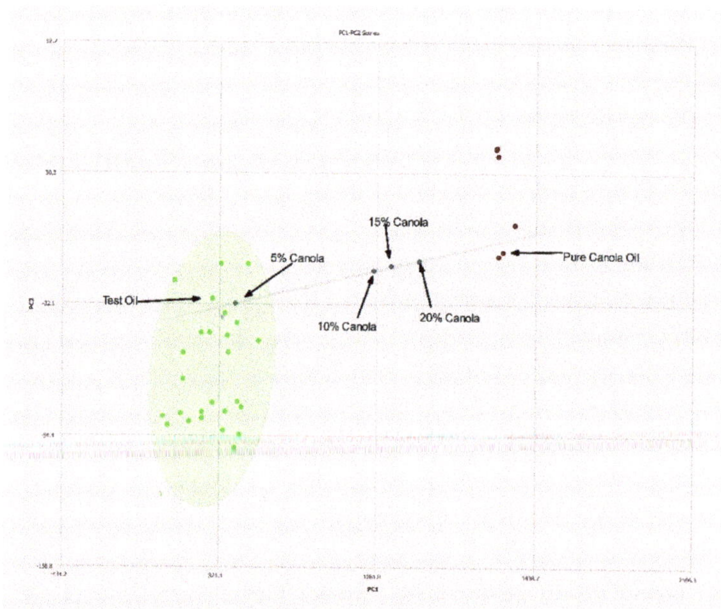

Figure 6. PC1–PC2 plot of 400 MHz ^{1}H NMR spectra of olive oil adulterated with canola oil with percentages of adulteration noted. Ellipses represent the 95% confidence interval.

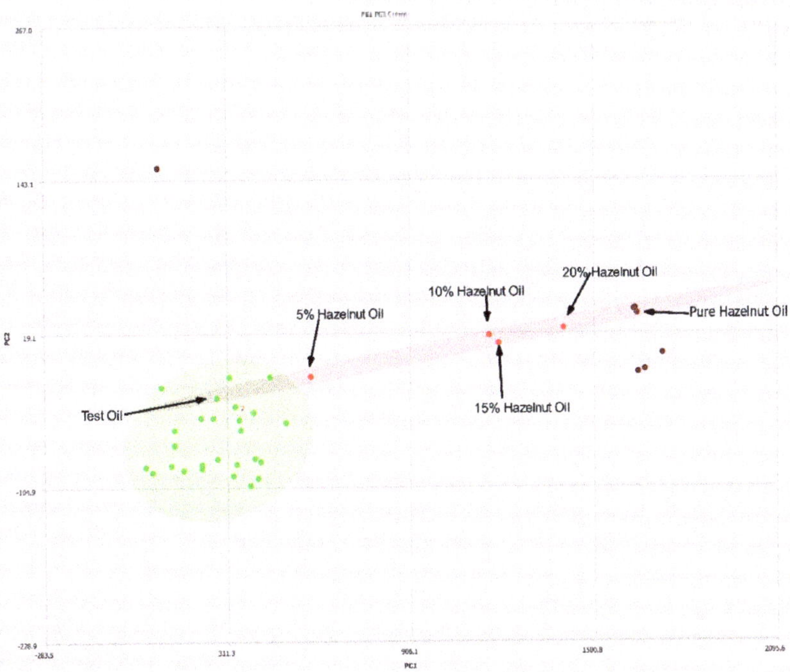

Figure 7. PC1–PC2 plot of 400 MHz ^{1}H NMR spectra of olive oil adulterated with hazelnut oil with percentages of adulteration noted. Ellipses represent the 95% confidence interval.

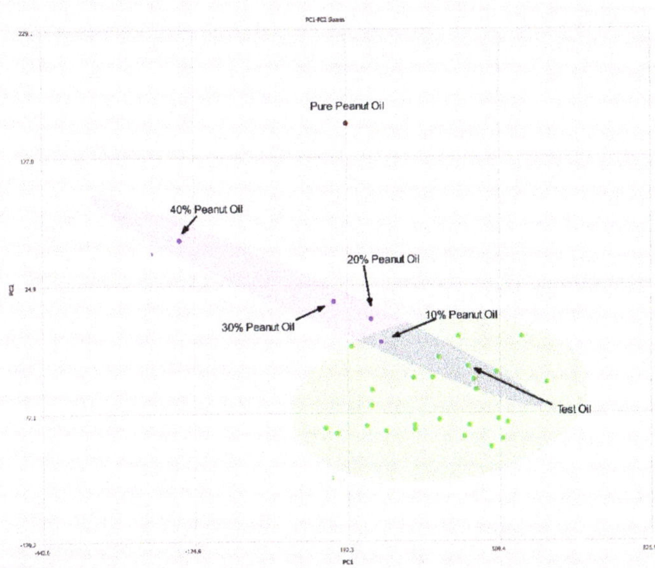

Figure 8. PC1–PC2 plot of 400 MHz ^{1}H NMR spectra of olive oil adulterated with peanut oil with percentages of adulteration noted. Ellipses represent the 95% confidence interval.

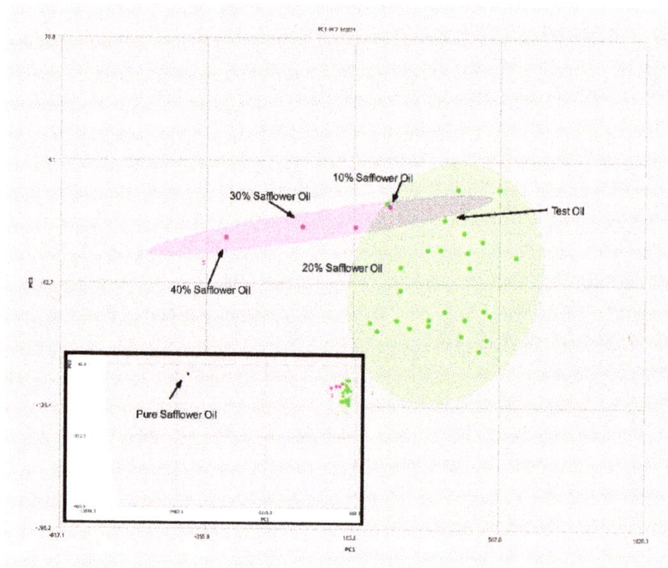

Figure 9. PC1–PC2 plot of 400 MHz ^{1}H NMR spectra of olive oil adulterated with safflower oil with percentages of adulteration noted. Ellipses represent the 95% confidence interval. The inset shows the expanded PCA plot to show the position of pure safflower oil with respect to the pure olive oil samples.

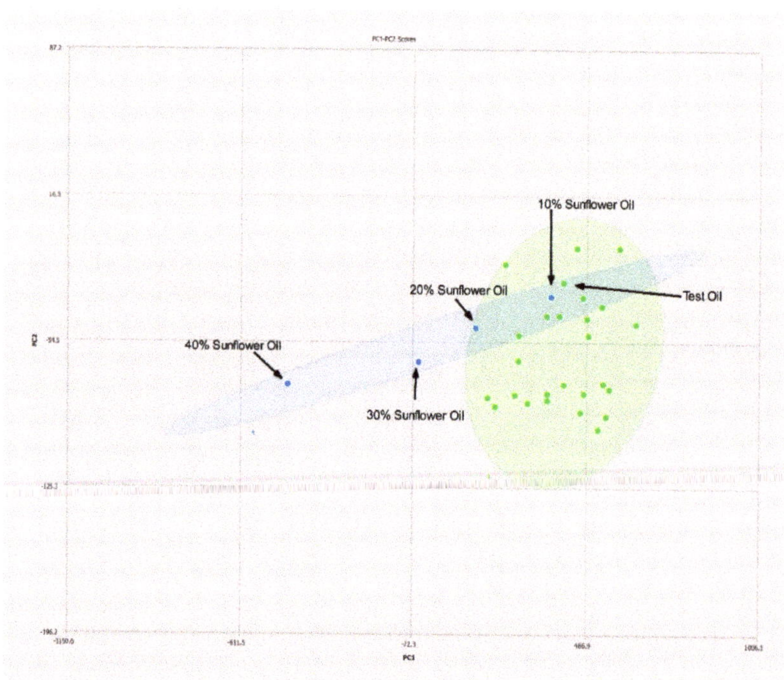

Figure 10. PC1–PC2 plot of 400 MHz 1H NMR spectra of olive oil adulterated with sunflower oil with percentages of adulteration noted. Ellipses represent the 95% confidence interval.

A commercially available blend of 70% canola, 20% grapeseed, and 10% olive oils was tested for the sake of comparison. As expected, the placement on the PC1–PC2 plot is nearest to canola oil with slight deviation toward both olive and grapeseed oils as seen in Figure 11.

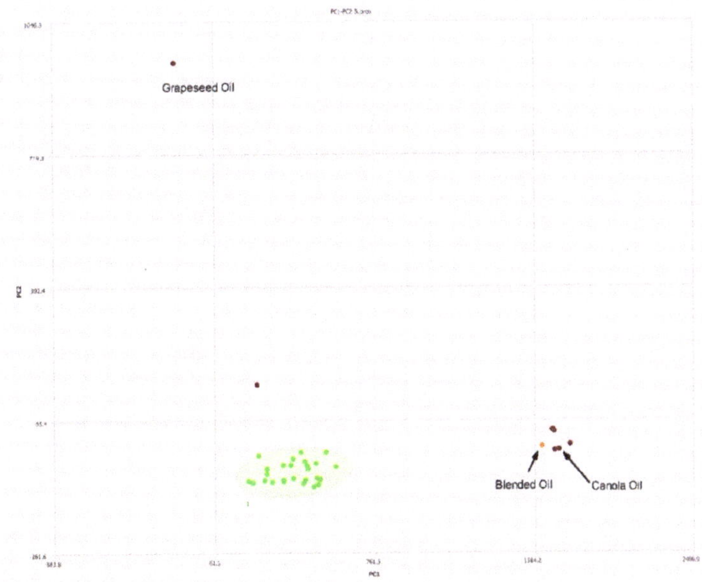

Figure 11. PC1–PC2 plot of 400 MHz ^{1}H NMR spectra of a commercially available blended oil composed of 70% canola, 20% grapeseed, and 10% olive oils (v/v). Ellipses represent the 95% confidence interval.

3.4. Detection of Adulterated Olive Oils

Upon successfully configuring this method it became clear that two of the olive oils, samples 1 and 27, originally purchased to determine a baseline for genuine olive oils appear to be adulterated themselves (Figure 11) as they lay outside the 95% confidence interval. Sample 1 was sold as a monovarietal arbequina extra virgin olive oil purchased from a from a boutique olive oil shop, and sample 27 was purchased from a grocer specializing in Middle Eastern products and was simply labeled "Moroccan Extra Virgin Olive Oil." Sample 1 was claimed to have been screened for authenticity via HPLC as part of grading. No testing was claimed for sample 27. Due to placement on the PC1-PC2 plot outside of the 95% confidence ellipse for olive oils, it is likely that both oils are adulterated at relatively low levels. Sample 1 is trending toward peanut or grapeseed oil. Sample 27 is trending toward a cluster of many other oils known to be used as adulterants and as such it is unclear as to which adulterant is included.

4. Discussion

Olive oils are known for having low levels of ω-3 fatty acids. By using the expected composition of olive oil during processing of oil spectra it is possible to greatly enhance the adulteration detection capability of a 400 MHz NMR. Typically, NMR spectra is normalized to the tallest peak in each spectrum. However, when one normalizes the intensity of oil spectra to the terminal methyl triplet of ω-3 fatty acids at 0.975 ppm as demonstrated here the spectral differences in non-olive oils are enhanced. As such adulterants such as high-oleic safflower and high-oleic sunflower oils that are ordinarily difficult to detect become readily detectable at economically viable levels of adulteration with a 400 MHz instrument. The data summarized in Figures 9 and 10 demonstrate the enhanced ability to discern olive oils adulterated with high oleic sunflower oil and safflower oil compared to traditional normalization techniques.

Adulteration of olive oils with low ω-3 fatty acid containing oils such as peanut, high-oleic safflower, high-oleic sunflower and high oleic canola oils are still quite easy to detect when using this method. This method does not specifically rely on the presence of sterols, terpines, phenolic compounds, or other low abundance marker compounds that are often destroyed or removed during refining, allowing this method to be used on refined olive oils, as well as virgin, or extra virgin oils.

While effective detection of olive oil adulteration remains a challenge, this method of leveraging processing techniques in order to determine the authenticity of olive oils quickly using more accessible NMR instrumentation should allow for wider availability of NMR-based authenticity testing. Even with only 26 genuine olive oils in this series of experiments, it was possible to detect two probable adulterated olive oils already with real world samples. The accuracy of the model will improve with greater numbers of genuine olive oil samples. The greatest benefit is the ability to screen these oils effectively with a 400 MHz NMR system without the need for a cryogenically cooled probe or specialized diffusion probe [10]. The significantly reduced cost of the equipment needed to perform these analyses and the higher throughput of NMR compared to chromatography-based analyses should make NMR a very competitive instrumentation choice for authenticity analysis purposes.

Author Contributions: Conceptualization, C.L.R. and C.M.G.; Data curation, J.A.G.; Project administration, C.M.G.; Writing—original draft, C.L.R.; Writing—review & editing, J.A.G. and C.M.G. All authors have read and agreed to the published version of the manuscript.

Funding: This research received no external funding.

Institutional Review Board Statement: Not applicable.

Informed Consent Statement: Not applicable.

Data Availability Statement: Data is contained within the article. The original data presented in this study are available on request.

Acknowledgments: We thank the University of Missouri-Columbia for financial support and research facilities used for this research.

Conflicts of Interest: The authors declare no conflict of interest.

Sample Availability: Samples of the compounds are available from the authors.

References

1. Yan, J.; Erasmus, S.W.; Toro, M.A.; Huang, H.; van Ruth, S.M. Food fraud: Assessing fraud vulnerability in the extra virgin olive oil supply chain. *Food Control.* **2019**, *111*, 107081. [CrossRef]
2. Boskou, D. *Olive Oil*, 2nd ed.; American Oil Chemists' Society: Urbana, IL, USA, 2006; pp. 41–72.
3. Castejón, D.; Fricke, P.; Cambero, M.I.; Herrera, A. Automatic 1H-NMR Screening of Fatty Acid Composition in Edible Oils. *Nutrients* **2016**, *8*, 93. [CrossRef] [PubMed]
4. Callaway, J.C. Hempseed as a nutritional resource: An overview. *Euphytica* **2004**, *140*, 65–72. [CrossRef]
5. Stoyanova, R.; Brown, T.R. NMR spectral quantitation by principal component analysis. *NMR Biomed.* **2001**, *14*, 271–277. [CrossRef] [PubMed]
6. Jabeur, H.; Zribi, A.; Makni, J.; Rebai, A.; Abdelhedi, R.; Bouaziz, M. Detection of Chemlali Extra-Virgin Olive Oil Adulteration Mixed with Soybean Oil, Corn Oil, and Sunflower Oil by Using GC and HPLC. *J. Agric. Food Chem.* **2014**, *62*, 4893–4904. [CrossRef]
7. Fang, G.; Goh, J.; Tay, M.; Lau, H.; Li, S. Characterization of Oils and Fats by 1H NMR and GC/MS Fingerprinting: Classification, Prediction and Detection of Adulteration. *Food Chem.* **2013**, *138*, 1461–1469. [CrossRef]
8. Mannina, L.; Sobolev, A.P. High resolution NMR characterization of olive oils in terms of quality, authenticity and geographical origin. *Magn. Reson. Chem.* **2011**, *49*, S3–S11. [CrossRef]

9. Alonso-Salces, R.M.; Moreno-Rojas, J.M.; Holland, M.V.; Reniero, F.; Guillou, C.; Héberger, K. Virgin Olive Oil Authentication by Multivariate Analyses of 1H NMR Fingerprints and δ13C and δ2H Data. *J. Agric. Food Chem.* **2010**, *58*, 5586–5596. [CrossRef] [PubMed]
10. Mejkalová, D.; Piccolo, A. High-Power Gradient Diffusion NMR Spectroscopy for the Rapid Assessment of Extra-Virgin Olive Oil Adulteration. *Food Chem.* **2010**, *118*, 153–158. [CrossRef]

molecules

MDPI

Article

Analytical Methods for the Characterization of Vegetable Oils

Agnese Giacomino [1], Paolo Inaudi [1,*], Gessica Silletta [1], Aleandro Diana [2], Stefano Bertinetti [2], Elisa Gaggero [2], Mery Malandrino [2], Federico Stilo [1] and Ornella Abollino [1]

[1] Department of Drug Science and Technology, University of Torino, 10125 Torino, Italy
[2] Department of Chemistry, University of Torino, 10125 Torino, Italy
* Correspondence: paolo.inaudi@unito.it

Abstract: The determination of the authenticity of extra virgin olive oils (EVOOs) has become more interesting in recent years. Italy is the first country in Europe in terms of number of Protected Designation of Origin (PDO) oils, which connects consumers to a feeling of tradition and thus to higher quality standards. This work focused on the consideration of the inorganic content as a possible marker of EVOOs. Ten vegetable oils (VOs), eight Italian EVOOs and seven not Italian EVOOs were analyzed. After pretreatment by acid mineralization, Al, Ba, Ca, Cu, Fe, K, Li, Mg, Mn, Na, P, Sb, Se and Zn were determined by ICP-OES. The electrochemical properties of a selected group of EVOOs and other vegetal oils of different botanical origin were investigated by voltammetry. Carbon paste electrodes (CPEs) were prepared. The features observed in the voltammograms reflect the reactions of electroactive compounds, which are present in the oils. A chemometric treatment of the results was performed to assess the possibility to distinguish (i) the region of provenience considering the inorganic profile; and (ii) the plant species from which each oil was obtained on the basis of the current profile registered during voltammetric analysis. Inorganic composition seems to be a useful marker for the assessment of the geographical origin of an EVOO. The EVOO-CPEs voltammetry seems to have a good ability to distinguish the plant species of origin. This method could be useful to monitor the conservation status of the oils, as the redox profile is linked to the oxidative degradation state.

Keywords: authentication; chemometric treatments; EVOO; marker; metals; redox profile

Citation: Giacomino, A.; Inaudi, P.; Silletta, G.; Diana, A.; Bertinetti, S.; Gaggero, E.; Malandrino, M.; Stilo, F.; Abollino, O. Analytical Methods for the Characterization of Vegetable Oils. *Molecules* **2023**, *28*, 153. https://doi.org/10.3390/molecules28010153

Academic Editor: Gavino Sanna

Received: 5 December 2022
Revised: 20 December 2022
Accepted: 21 December 2022
Published: 24 December 2022

1. Introduction

In the context of the global market, the certification of food quality is one of the most important goals for scientists in the agri-food sector. The determination of the authenticity and traceability of food products is of great interest for consumers, producers and distributors, and for these reasons international agencies have published specific guidelines on quality standards for oils [1].

Vegetable oils (VOs) and extra virgin olive oils (EVOOs) are widely used in the cooking and alimentary, cosmetic, pharmaceutical and chemical industries [2]. The EVOO is considered essential for the Mediterranean diet, and for this reason it has been extensively investigated in an effort to identify its geographical origin and detect frauds and adulteration [3–6].

The European Union (EU) has addressed the regulation, production and commercialization of oils. EU Regulation No. 1019/02 defines how to correctly pack and label oils, and the Commission Implementing Regulation EU No. 1335/13 (2013) made it obligatory to indicate the geographic origin on the label. According to EU Regulation No. 29/12 (European Commission Implementing Regulation, 2012), information concerning the geographic area in which olives are harvested and olive oil is obtained should be stated on the packaging or the label. For greater clarity, the document also defines that simple provisions such as 'blend of olive oils of European Union origin' or 'blend of olive oils not of European Union origin' or 'blend of olive oils of European Union origin and not of European Union origin' should be stated for the labeling of the origin of the oils.

Olive trees are widely cultivated throughout the Mediterranean area and give rise to products of prominent peculiarities and quality when compared with products in other parts of the world. The European Union is the leading producer, consumer and exporter of olive oil. On average, the EU has produced 68.4%, consumed 54.2%, and exported 66.9% of the world's olive oil since 2013 [7]. In particular, Italy plays an important role in the production of high-quality oil [8], also differentiated by region of origin. Specifically, more fruity flavors and a fairly pronounced taste are associated with oils of the northern part of Italy, while a fuller and more strong character is more typical of ones from the south [9].

The olive tree belongs to the *Oleaceae* family, and thus it is an evergreen tree. It is a plant typical of the Mediterranean flora, but is able to adapt to temperate climates at other latitudes as well. *Olea Europaea L.* has many subspecies, however the distinction of particular economic importance is the one between the wild varieties (*Olea Europaea Var. Sylvestris (Mill.) lehr*), characterized by small fruits containing little oil, and the cultivated variety (*Olea Europaea Var. Europaea L.*), which produces edible fruits with characteristics that vary greatly depending on the cultivar, the soil, and the climate conditions in which the plant has grown.

Due to the physiological activity of the cells during the ripening of the fruit, which can occur in different periods of the year depending on the cultivar, the chemical composition of the drupe may be modified both for the organic fraction (fatty acids and minor components) and the inorganic (e.g., more magnesium and phosphorus, less calcium) one. With maturation, in fact, the number of water-soluble substances is reduced in favor of the lipid component: this process implies an increasingly critical state for the preservation of olive oil and consequently for its future quality. The oil is, in fact, predominantly contained in vacuolar structures within the cell, well separated from the cytoplasmic enzymes. In cases of excessive maturation, vacuolar barriers degrade and allow the contact of lipase and lipoxidase with the lipid content of these vacuoles. Due to the activity of these two enzymes, the inactivation of triglycerides (by action of lipase) and rancidity (due to the intervention of lipoxidase) may occur [10,11].

For this reason, the time of harvest of olives greatly influences the quality of the oil produced: it is customary to consider an oil obtained from overripe olives to be of inferior quality, whereas olives that are too immature will have a more aggressive taste of bitterness and spiciness [12].

Adulteration is a serious issue for EVOOs, and it may take place by mixing EVOO with oils of lower quality, for example by soybean oil [13]. Many analytical techniques have been used to assess the quality and investigate the adulteration of oils, for example 1H and 13C nuclear magnetic resonance (NMR) imaging [14], near infrared resonance (NIR) [15] and mid infrared resonance (MIR). Some other spectroscopic techniques, such as Raman spectroscopy [16–18], fluorescence spectroscopy [19–21], Fourier transform infrared (FTIR) spectroscopy [22,23], and ultraviolet spectrometry [24,25] have great potential for characterizing different olive oil samples from the point of view of organic substances and hence detecting their adulteration. Inductive coupled plasma optical emission spectroscopy (ICP OES), inductive coupled plasma mass spectrometry (ICP MS) and graphite furnace atomic absorption spectrometry (GF AAS) [26–30] are widely used to investigate the trace element content in oils.

In addition, high performance liquid chromatography (HPLC) and gas chromatography (GC) [31–37], frequently coupled with MS, were employed to detect EVOOs adulteration concerning organic components. Several publications have described the use of volatile-species distribution as a fingerprint to assess traceability, authentication, and non-degradation based on head-space sampling and GC. Most of these techniques are expensive and/or complex; thus, a simpler, more rapid, and cost-effective procedure [38] is highly desirable. Voltammetry fulfils these requirements and may represent a suitable alternative for detecting adulteration in olive oil. In two works, Zappi et al. [39,40] suggested that correlations between the electrochemical response of a modified screen-printed electrode and the olives cultivar and/or their geographical origin could be established.

The objective of the present manuscript are twofold. The first one is to investigate the relationship between the inorganic components of oils, determined by ICP-OES, and their geographical origin, matching the definition of an "inorganic fingerprint" characterizing each oil. Metals present in oils mainly derive from soils in which olive trees were grown, but may have other sources: atmospheric fallout, fertilizers, and metal-containing pesticides, or contamination from the metal-based processing equipment [27,41].

The second aim is to investigate the oxidation and reduction profile of different oils [42] related to the presence of antioxidant compounds, such as polyphenols, tocopherols, carotenoids and chlorophylls using voltammetry. Analyses were performed both by cyclic voltammetry (CV) and square wave voltammetry (SWV) using modified carbon paste electrodes (CPE) in which the paste binder was made of the oil under consideration. The purpose is to examine the possibility of discriminating against the botanical origin of the different oils examined based on their content in terms of antioxidant compounds.

The results of chemical analysis were processed with principal components analysis (PCA) [43] to search for a fingerprint of origin of oils, resulting from the trace element determination and voltammetric analysis.

2. Materials and Methods

2.1. Samples

A total of 15 EVOO samples were investigated. Table 1 lists their codes, origins and classification: among these samples there are some monovarietal productions (indicated with MV), some Protected Designation of Origin (PDO) specimens, and some blends (mix of olives of different varieties). Eight samples come from two different regions of Italy: two from Liguria and six from Umbria. All the samples analyzed were produced and harvested in 2018–2019. Another seven oils of non-Italian production were analyzed: one from Croatia, four from Spain, and two from Portugal.

Table 1. List of EVOOs samples.

Code	Region	Classification
EVOO1	Liguria	MV
EVOO2	Liguria	MV
EVOO3	Umbria	PDO
EVOO4	Umbria	PDO
EVOO5	Umbria	Blend
EVOO6	Umbria	Blend
EVOO7	Umbria	Blend
EVOO8	Umbria	Blend
EVOO9	Croatian	MV
EVOO10	Spain	PDO
EVOO11	Spain	MV
EVOO12	Spain	MV
EVOO13	Spain	MV
EVOO14	Portugal	PDO
EVOO15	Portugal	PDO

MV: monovarietal, PDO: Protected Designation of Origin, Blend: mix of olives of different varieties.

Ten samples of other vegetable oils (VOs), namely sunflower seed oil, corn seed oil, peanut seed oil, almond oil and sesame seed oil, purchased from a supermarket, have also been analyzed to compare their composition, in terms of redox properties and inorganic elements, with the composition of EVOOs. Samples are listed in Table 2.

Table 2. VO samples.

Code	Classification
SS1	Sunflower Seed Oil (Brand A)
SS2	Sunflower Seed Oil (Brand B)
SS3	Sunflower Seed Oil (Brand C)
CS1	Corn Seed Oil (Brand A)
CS2	Corn Seed Oil (Brand B)
CS3	Corn Seed Oil (Brand D)
PS1	Peanut Seed Oil (Brand A)
PS2	Peanut Seed Oil (Brand E)
A	Almond Oil (Brand F)
SeS	Sesame Seed Oil (Brand G)

2.2. Apparatus and Reagents

Sample dissolution was performed in tetrafluoromethoxyl (TFM) vessels, with a Milestone MLS-1200 Mega (Milestone, Sorisole, Italy) microwave laboratory unit.

Analyses of element contents were carried out with an inductively coupled plasma-optical emission spectrometer (ICP-OES), in particular the Perkin Elmer Optima 7000 (Perkin Elmer, Norwalk, CT, USA).

Analytical grade reagents were used throughout. Standard metal solutions were prepared from concentrated Merck TraceCERT stock solutions (Merck, Darmstadt, Germany).

Electrochemical analyses were made with an Autolab PGSTAT 10 analyzer (Eco-Chemie Metrohm, Utrecht, The Netherlands) connected to a Metrohm 663 VA Stand. The potentiostat is connected to a computer that allows the setting of conditions measurement, the display of voltammograms and the related data through the IME 663 interface and the GPES software (General Purpose Electrochemical System).

High purity water (HPW) obtained from a Milli-Q apparatus (Millipore, Bedford, MA, USA) was used throughout for the preparation of sample and standard solutions.

2.3. Procedures

2.3.1. Profile of Inorganic Component

Unfortunately, no Standard Reference Materials (SRM) for trace elements in oils is available on the market. For this reason, SRM 1573a, tomato leaves, supplied by the National Institute of Standards and Technology (NIST, Gaithersburg, Maryland, United State) were analyzed to value the efficiency and the accuracy of analytes quantification. This SRM is primarily intended for use in botanical materials, agricultural food products, and materials of similar matrix.

It is very important to homogenize oil samples before the pretreatment, since the oil matrix tends to stratify, especially when there is a residue at the bottom of the container and/or if the oil has been subjected to thermal stress (stored in the freezer and then thawed for analysis). Acid digestion in the microwave oven was adopted to mineralize the samples, according to our previous work on similar matrices [44,45]. The stirring step is indispensable to homogenize each sample.

Aliquots of 0.5 g of each sample were weighed directly in TFM vessels, then 3 mL of HNO_3 65% and 3 mL of H_2O_2 30% were added. The digestion sequence (1 min–250 W, 1 min–0 W, 5 min–250 W, 5 min–400 W, 5 min–600 W, 20 min of ventilation) was applied through an external microprocessor, and then the vessels were left under a fume hood for two hours for a final cooling phase.

The sample solutions so obtained were analyzed by ICP-OES for a quantitative evaluation of Al, Ba, Ca, Cu, Fe, K, Li, Mg, Mn, Na, P, Sb, Se and Zn. For each analyte the optimal wavelength for signal measurement was chosen to obtain maximum response sensitivity and at the same time minimum interference by other species. The quantification of the elements was carried out following the external calibration method.

2.3.2. Reductive Oxide Profile

CPEs were prepared by mixing graphite powder and the oil under investigation. Graphite and oil were previously weighed and homogenized. The obtained paste was then packed in the body of the electrode, represented by a polypropylene tube obtained from a syringe, after cutting its tip, and an electric contact, namely a copper wire, was inserted. The syringe plunger was put in contact with the paste, after removing its tip, and was used as a piston to remove the outer part of the paste after the analysis. Figure 1a–e shows the structure of the CPE.

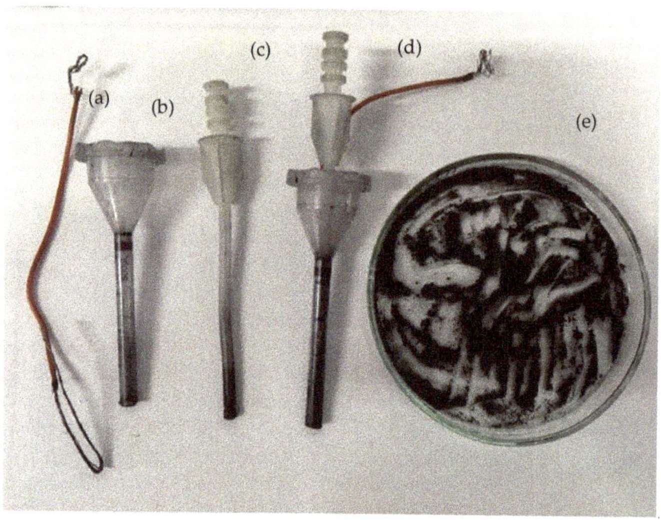

Figure 1. Structure of the CPE (**a**) copper wire, (**b**) polypropylene tube, (**c**) syringe pluger, (**d**) complete CPE, (**e**) paste/oil mix.

After packing, the paste was smoothed by rubbing its outer surface on a sheet of weighing paper. Generally, the CPEs so obtained were used as prepared.

For each test sample, more than one CPE was prepared by varying the ratio between graphite powder and oil: 80:20, 70:30, 60:40 and 50:50, respectively. Each CPE was tested as a working electrode (WE) in an electrochemical cell filled with 0.1 M HCl as the supporting electrolyte.

The best choice for the analysis was found to be the graphite/binder mixture in a 70:30 ratio, since both the voltammograms obtained by cyclic voltammetry (CV) and square wave voltammetry (SWV) show the most defined peaks and have background currents close to zero. Thus, this ratio was used throughout the study.

All samples were analyzed by CV. The scanning parameters set for the analysis were: potential range: 0–1, 15–0 V, scan: linear scan, scan rate: 0.1 V/s.

For all samples, a SWV analysis was also performed, scanning the potential in anodic and cathodic direction. This type of scan, using a pulsed waveform, generally allows for the obtaining of more intense signals and consequently more defined peaks than a linear scan. The parameters set in this case were: potential range: 0–1.15 V for oxidation, 1.15–0 V for reduction, potential step: 0.007 V, wavelength amplitude: 0.010 V, frequency: 15 Hz.

2.4. Chemometric Treatment

A PCA was carried out with the aid of the XLSTAT4.4 software package (Addinsoft, Paris, France), a Microsoft Excel plug-in. Unscrambler X 10.2 (Camo Analytics, Oslo, Norway) was employed for auto-scaling the dataset and for substituting values below the limits of detection (LODs) with estimated values.

3. Results and Discussion

3.1. Inorganic Component

3.1.1. Reference Material

Table 3 reports the results concerning the analysis of SRM 1573a, tomato leaves. Percentage recoveries from 75% to 101.5% were obtained.

Table 3. Results of the analysis of SRM—Tomato Leaves (mg kg^{-1}).

Element	Certified Value	Experimental Result	Recovery (%)
Al	598 ± 12	451 ± 15	75.5
B	33.3 ± 0.7	27.5 ± 0.3	82.7
Ca	50,500 ± 900	45,072 ± 1604	89.2
Cd	1.52 ± 0.04	1.32 ± 0.01	87.1
Co	0.57 ± 0.02	0.64 ± 0.01	112
Cr	1.99 ± 0.06	1.87 ± 0.02	94.2
Cu	4.70 ± 0.14	4.05 ± 0.09	86.2
Fe	368 ± 7	284 ± 4	77.2
K	27,000 ± 500	24,455 ± 916	90.6
Mg	12000 *	9706 ± 307	80.9
Mn	246 ± 8	197 ± 5	80.1
P	2160 ± 40	2192 ± 28	101.5
Sr	85 *	67.1 ± 0.8	79.0
Zn	30.9 ± 0.7	27.0 ± 0.4	87.4

*informative value.

The accuracy of the results can be considered satisfactory: for this reason, the same procedure was applied to the oil samples.

3.1.2. EVOOs and other Vegetable Oil Samples

Table 4 a and b show the results of the analysis of the inorganic components in oil samples. The data show that Al is present in all samples in concentrations of 0.6–0.9 mg kg^{-1}, with the exception of EVOO1, EVOO13, almonds, and sesame oils in which it exceeds 1 mg kg^{-1}. Al is considered toxic for plants at concentrations between 2 and 5 mg kg^{-1}, since it limits their development [46]. Ca is an essential element; it is the element with the highest concentration in all samples. Ca content in EVOOs varied from 4 to 16 mg kg^{-1}, while higher concentrations are present in several vegetable oils: in particular, in sesame oil and in almond oil its concentration is 29 mg kg^{-1} and 37 mg kg^{-1}, respectively. The concentration range for Fe is 0.45–1.18 mg kg^{-1} in the Italian and Croatian EVOOs, while it is below the detection limit in the other EVOOs and in most vegetal oils. K shows a high variability of concentration in EVOOs (0.06–17.4 mg kg^{-1}); it is mainly present in Umbrian and Croatian EVOOs, where it exceeds 5 mg kg^{-1}. Mg and Na are present in most samples at comparable concentrations of 0.9–3 mg kg^{-1}, with the exception of a few samples with lower Na concentrations and SES in which 9.32 and 4.46 mg kg^{-1} of Mg and Na are found, respectively. Se is a trace element found in all samples, albeit in very variable concentrations, irrespective of the plant species of origin. Some elements, namely Ba, Cu, Li, Mn, P, Sb and Zn, are not reported, because they were always below the LOD of the method.

Table 4. Inorganic element content (mg kg^{-1}) in (a) EVOOs and (b) VOs.

EVOO							
Sample	**Al**	**Ca**	**Fe**	**K**	**Mg**	**Na**	**Se**
EVOO 1	1.22 ± 0.03	11.4 ± 0.3	0.59 ± 0.02	0.43 ± 0.06	2.90 ± 0.04	2.59 ± 0.09	1.42 ± 0.03
EVOO 2	0.79 ± 0.02	9.32 ± 0.30	0.99 ± 0.02	9.97 ± 0.72	2.08 ± 0.02	2.80 ± 0.07	1.47 ± 0.02
EVOO 3	0.65 ± 0.02	16.2 ± 0.4	0.94 ± 0.02	5.53 ± 0.15	1.86 ± 0.02	1.07 ± 0.08	4.17 ± 0.02
EVOO 4	0.68 ± 0.02	5.59 ± 0.18	0.56 ± 0.03	15.9 ± 0.02	2.11 ± 0.11	1.69 ± 0.10	1.25 ± 0.08
EVOO 5	0.67 ± 0.02	4.72 ± 0.19	0.70 ± 0.02	17.4 ± 0.5	1.56 ± 0.02	1.75 ± 0.07	0.95 ± 0.06
EVOO 6	0.94 ± 0.02	8.36 ± 0.61	0.47 ± 0.02	9.91 ± 0.13	2.26 ± 0.03	1.44 ± 0.11	1.10 ± 0.02
EVOO 7	0.83 ± 0.02	6.99 ± 0.10	0.49 ± 0.02	11.5 ± 0.2	2.17 ± 0.05	1.00 ± 0.02	1.78 ± 0.11
EVOO 8	0.60 ± 0.02	4.02 ± 0.18	0.89 ± 0.02	8.55 ± 0.22	1.47 ± 0.02	1.15 ± 0.06	2.21 ± 0.05
EVOO 9	0.75 ± 0.02	7.74 ± 0.60	1.18 ± 0.02	8.33 ± 0.29	2.07 ± 0.07	1.91 ± 0.03	1.43 ± 0.03
EVOO 10	0.60 ± 0.02	14.6 ± 0.3	<0.02	0.06 ± 0.06	1.54 ± 0.02	0.90 ± 0.15	3.61 ± 0.10
EVOO 11	0.70 ± 0.02	9.78 ± 0.13	<0.02	2.02 ± 0.05	1.90 ± 0.06	1.94 ± 0.02	1.16 ± 0.08
EVOO 12	0.72 ± 0.02	11.1 ± 0.3	<0.02	1.07 ± 0.03	2.03 ± 0.02	1.86 ± 0.17	1.09 ± 0.11
EVOO 13	1.21 ± 0.03	13.3 ± 0.7	<0.02	0.89 ± 0.06	2.75 ± 0.02	1.43 ± 0.04	1.08 ± 0.29
EVOO 14	0.63 ± 0.02	8.66 ± 0.30	<0.02	0.35 ± 0.02	1.82 ± 0.03	1.38 ± 0.28	1.90 ± 0.17
EVOO 15	0.64 ± 0.02	5.98 ± 0.21	<0.02	0.42 ± 0.02	1.95 ± 0.02	1.40 ± 0.08	1.79 ± 0.02
MIN	0.60	4.02	<0.02	0.06	1.47	0.90	0.95
MAX	1.22	16.2	1.18	17.4	2.90	2.80	4.17
MEAN	0.77	9.19	0.46	6.15	2.03	1.62	1.76

VO							
Sample	**Al**	**Ca**	**Fe**	**K**	**Mg**	**Na**	**Se**
CS1	0.60 ± 0.02	6.73 ± 0.18	<0.02	0.71 ± 0.02	2.22 ± 0.02	1.61 ± 0.19	1.58 ± 0.03
CS2	0.61 ± 0.03	13.9 ± 0.5	<0.02	<0.02	1.39 ± 0.02	0.24 ± 0.04	2.00 ± 0.08
CS3	0.66 ± 0.02	9.91 ± 0.81	<0.02	<0.02	1.15 ± 0.02	<0.02	1.61 ± 0.08
SS1	0.55 ± 0.02	12.3 ± 0.1	<0.02	0.03 ± 0.02	1.55 ± 0.07	0.24 ± 0.18	1.36 ± 0.02
SS2	0.64 ± 0.02	12.1 ± 0.5	<0.02	<0.02	1.56 ± 0.05	0.79 ± 0.08	1.17 ± 0.13
SS3	0.89 ± 0.02	8.25 ± 0.41	2.72 ± 0.03	8.26 ± 0.02	2.30 ± 0.03	1.91 ± 0.13	2.36 ± 0.05
PS1	0.68 ± 0.02	20.3 ± 0.7	< 0.02	<0.02	1.77 ± 0.02	0.45 ± 0.08	3.44 ± 0.10
PS2	0.84 ± 0.02	11.7 ± 0.6	2.45 ± 0.02	5.17 ± 0.09	2.15 ± 0.05	1.38 ± 0.24	2.13 ± 0.04
A	1.09 ± 0.02	37.0 ± 1.0	<0.02	0.49 ± 0.06	2.67 ± 0.05	1.23 ± 0.02	0.83 ± 0.10
SeS	1.54 ± 0.02	29.0 ± 0.6	2.67 ± 0.00	8.00 ± 0.30	9.32 ± 0.20	4.67 ± 0.05	1.07 ± 0.02
MIN	0.55	6.73	<0.02	< 0.02	1.15	<0.02	0.83
MAX	1.54	37.0	2.72	8.26	9.32	4.67	3.44
MEAN	0.81	16.12	0.79	2.12	2.61	1.25	1.75

It can be concluded that the presence of essential elements such as Na, Mg and Ca was detected in all samples, regardless of the botanical origin of the oil.

Some differences can be observed: EVOOs from Liguria contain a higher content of Al and Na than Umbrian EVOOs, but they present lower concentrations of K; the amount of Fe is comparable. It can be assumed that this is due to the peculiarities of the growing area.

Comparing Italian and non-Italian EVOOs, Fe is present at detectable levels in the former, while it is below the LOD in the latter, with the exception of Croatian EVOO that is produced on the Adriatic coast.

Sesame oil generally presents the highest concentration of most of the elements, namely Al, K, Mg and Na, and also has a high Ca level. Almond oil presents a remarkably high content of Ca.

The metal content mainly depends on the conditions in which the olives were grown, namely the composition of soil and water. In particular, the differences are influenced by the pedoclimatic conditions of growth of the botanical species of origin. For example, Umbria is a region of central Italy surrounded by green hills and mountains, while Liguria is a northern region characterized by the presence of coasts for half of its borders; the relatively high concentrations of Na in Ligurian samples, in comparison with the other

EVOOs, may be due to the contribution of marine spray. Furthermore, the conditions of production, storage and transport of oil may affect the element content [26].

3.2. Redox Profiles

3.2.1. Method Development with Commercial EVOO

The redox profile of the samples was investigated. It is related to the presence of antioxidants in the oil matrix, capable of taking part in electron transfers typical of redox reactions.

Each oil was mixed with the graphite powder, acting simultaneously as both binder and sample. The electrochemical response of each CPE reflects the oxide-reductive properties of electroactive compounds present in it, such as polyphenols, tocopherols, carotenoids and chlorophylls. CPEs were prepared just before use. In any case, a repeatability of the response was verified after three days, during which they were stored at room temperature in the dark and covered with Parafilm.

Figures 2 and 3 show the voltammograms obtained by CV and SWV, respectively, using different oil: graphite ratios (20:80, 30:70, 40:60 and 50:50). Experiments were performed with an EVOO purchased in a local supermarket.

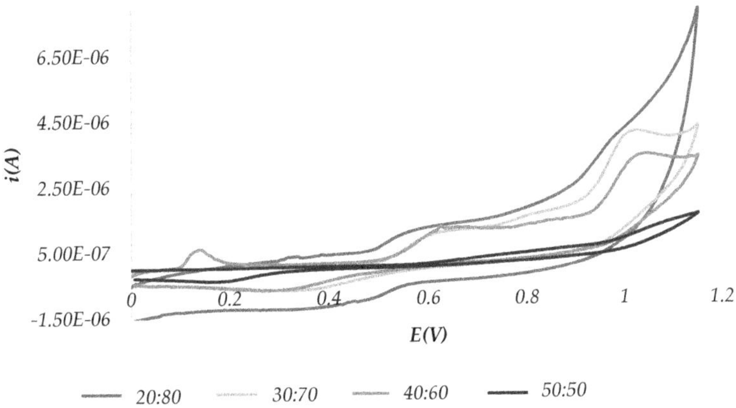

Figure 2. CV scans recorded, varying the oil: graphite ratio in the CPE.

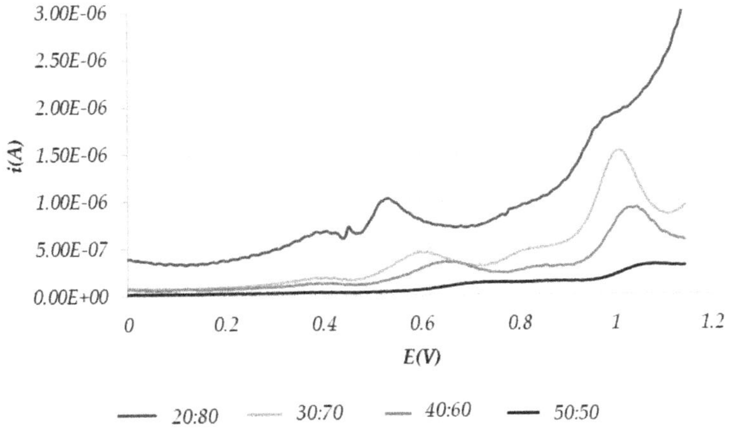

Figure 3. SWV scans recorded varying the oil:graphite ratio in the CPE.

Overall, the most defined peaks are shown, both in CV and SWV, using the graphite/oil 70:30 ratio, so this ratio was used throughout the study.

In accordance with the literature [47], the observed peaks can be traced back to a given electroactive chemical family. Namely:

- peaks around the potentials of 0.4 V and 0.6 V correspond to polyphenols
- the peaks around the 1.1 V potential are relative to tocopherols.

3.2.2. EVOO and VO Samples

The voltammograms acquired with SWV show more defined peaks than those obtained using CV, but the results of both techniques will be discussed hereafter, because they provide complimentary information.

In EVOOs (Supplementary Materials Figures S1–S4) the redox profiles obtained with anodic scans show up to four peaks: as stated above, the peaks correspond to two chemical species, namely polyphenols at potential around 0.4 V, 0.6 V and 0.8 V and tocopherols at about 1 V.

Cathodic and CV scans (Supplementary Materials Figures S5–S12) display from two to three peaks in the voltammograms relating to the EVOOs and from one to two peaks in those associated with other oils (Supplementary Materials Figures S17–S20). Specifically, for EVOOs there are one or two peaks between 0.4 V–0.6 V, and in more than half of the samples there is also a peak at around 1.1 V.

In VOs (Supplementary Materials Figures S13–S24) the peaks related to SWV-A are less numerous (up to three in Supplementary Materials SS2 and SS3, Figure S13) and have lower height than in EVOOs. In all cases there is a low peak at 0.4 V (polyphenols), in sunflower and corn there are peaks around 0.6 V (polyphenols, Supplementary Materials Figures S13–S14) and in Supplementary Materials SS2 and SS3 there is a third peak around 1.1 V (tocopherols, Supplementary Materials Figure S17).

In VOs, instead, there is always a peak at 0.4 V and in some samples a second peak appears: for CS1, CS2, PS2 (Supplementary Materials Figures S14–S15) and all corn seed oils it is found at about 1.1 V, while for sesame oil it is observed near 0.2 V (Supplementary Materials Figures S16 and S20).

It is possible to notice some differences between the profiles of the EVOOs and those of VOs. The first one shows defined and higher peaks for tocopherols at potentials of about 0.6 and 1 V, and in some cases a small peak at 0.4 V. In the voltammograms of other oils it is possible to recognize a less sharp peak at about 0.6 V for sunflower, corn and peanut, and at about 0.4 V for sesame and almond. From this first comparison we can already assert that a peak related to polyphenols is present in all oils, independently of the botanical species of origin, while the tocopherols seem most present in the EVOOs.

The results obtained by both CV and SWV suggest that EVOOs are richer in redox-active species, such as polyphenols and tocopherols, compared to VOs; therefore voltammetry confirms that the former represent a valuable source of antioxidants.

Peaks are different in shape, position or intensity, being associated with antioxidant molecules of different natures and present in different concentrations in relation to the botanical species from which the oil comes.

3.2.3. Applications of Redox Profiles

Identification of Possible Adulterations with Other Oils

To assess the applicability of the technique to check the presence of counterfeits, due to the adulteration of EVOOs with oils of different botanical origin, mixtures of EVOO and other vegetable oils in a ratio of 5:1 were prepared. CPEs were made with these mixtures, and SWV-A voltammograms were compared with those of the two oils of origin.

Figure 4 shows, as an example, the comparison between the voltammograms obtained with EVOO1 (100%): CPE, SS1 (100%)-CPE and EVOO1:SS1 = 5:1 mixture-CPE.

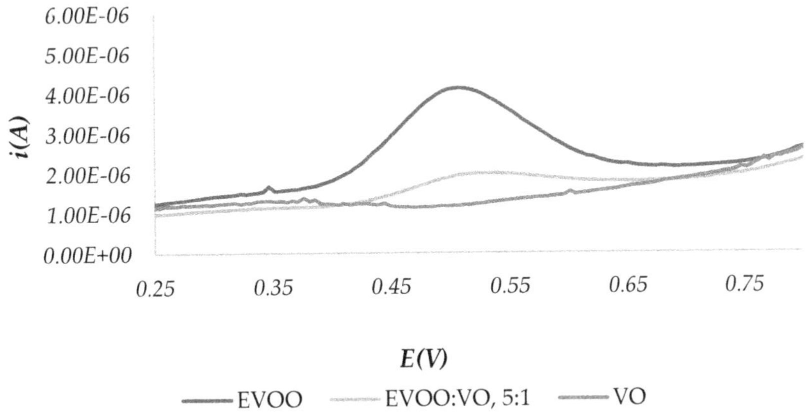

Figure 4. SWV-A scans for the comparison of a mixture of EVOO and SSO with the pure oils.

It is possible to observe that a small amount of SS significantly changes the redox profile, thanks to the variation of the amount of the antioxidant species characteristic of each oil. These results show that the comparison of voltammograms can be used to detect the presence of an adulteration of EVOOs.

Assessment of the Conservation Status

An improper storage of oils can negatively affect their stability and consequently give rise to changes in chemical composition, which in turn can affect their organoleptic characteristics. To assess the applicability of the technique to monitor the conservation status of the oil, an aliquot of EVOO was exposed to direct sunlight in a closed vessel. Three CPEs were prepared: the first one using a sample stored in the dark (hereafter called "fresh EVOO"), and the second and third with the sample after one and two months of exposure, respectively. SWV-A voltammograms were recorded. The results are reported in Figure 5.

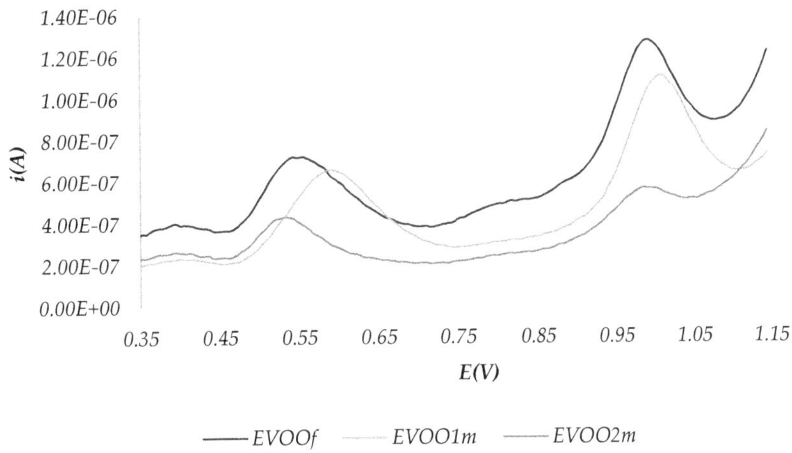

Figure 5. SWV recorded for fresh EVOO (EVOOf), after one (EVOO1m) and two (EVOO2m) months of exposure to sunlight.

It is possible to observe a difference between the three samples. In particular, the peaks at 0.5 and 1.0 V decrease after two months of exposure to sunlight. The oxide-reduction profile could therefore provide informative data on the state of each oil in terms of both proper preservation, especially for relatively long periods, and organoleptic quality.

3.3. Chemometric Treatment

The experimental results were processed by Principal Component Analysis (PCA), one of the main multivariate chemometric techniques, which enables a multi-variable reality to be represented with a few variables, namely the PCs. The latter are obtained by linear combinations of the original variables and are orthogonal to each other. PCs maintain the total variance of the original variables but change its distribution: most of the variance is retained in the first two or three PCs. PCA allows to reduce the size of data, represent experimental data in an orthogonal space, eliminate spurious information such as instrumental background noise, assess the relative relevance of the variables, and visualize the samples graphically allowing for the search for classes, clusters and outliers. PCA plots rely on two graphical elements: scores, i.e., points representing samples in analysis; and loadings, i.e., vectors with origin at the intersection of the axes representing the variables. PCA loadings are the coefficients of the linear combination of the original variables from which the principal components (PCs) are constructed. These elements can be represented individually (loading and score plots) or in combination (biplots).

In this way, it is possible to evaluate the degree of similarity of the different samples on the basis of investigated variables, which in this work are the concentrations of the elements and the current intensities measured during the potential scans.

3.3.1. PCA of the Inorganic Components

The biplot for all of the oils (Figure 6A) shows how oils with similar origin are closer together in the graphic representation. Concerning EVOOs (Figure 6B), Portuguese EVOOs (highlighted in blue in the Figure 6B) are very close to each other, their great similarity deriving from the fact that they were supplied by the same producer that cultivates in the same area; they are characterized by the important content of Se. The other EVOOs tend to be grouped in a less defined area. The Spanish EVOOs are grouped in the same quadrant, in correspondence to relatively high concentrations of Ca, Al and Mg, with the exception of EVOO10, which is closer to the Portuguese samples. This behavior could be explained by the fact that it has a greater concentration of Se and that it is a single variety PDO (like the Portuguese), while the other Spanish samples are made from blends of different varieties of olive. The Umbrian EVOOs are also quite close to one another, around the loadings of K and Fe, but occupy a much larger area in the plot; an exception is given to EVOO3, which is a bit further away from all other samples owing to the greater quantity of Se. The Ligurian EVOOs, on the other hand, are both towards the right end of the chart, although not being adjacent to each other, because of the different quantities of Al and Mg. It should be noted that the Italian EVOOs are in the upper part of the chart, except for EVOO3 and EVOO1, which, however, are very close to the axis, so they are characterized by a greater presence of K, Fe and Na than the other EVOOs. Furthermore, EVOO9 of Croatia is located in the upper part of the plot; it is possible to assume that this is due to its geographical origin, since the company that produces it is located on the Adriatic coast, a few dozen kilometers away from the Italian one. On the contrary, non-Italian EVOOs, being at the bottom of the graph, are characterized by a greater similarity related to the concentrations of Se, Ca, Al and Mg.

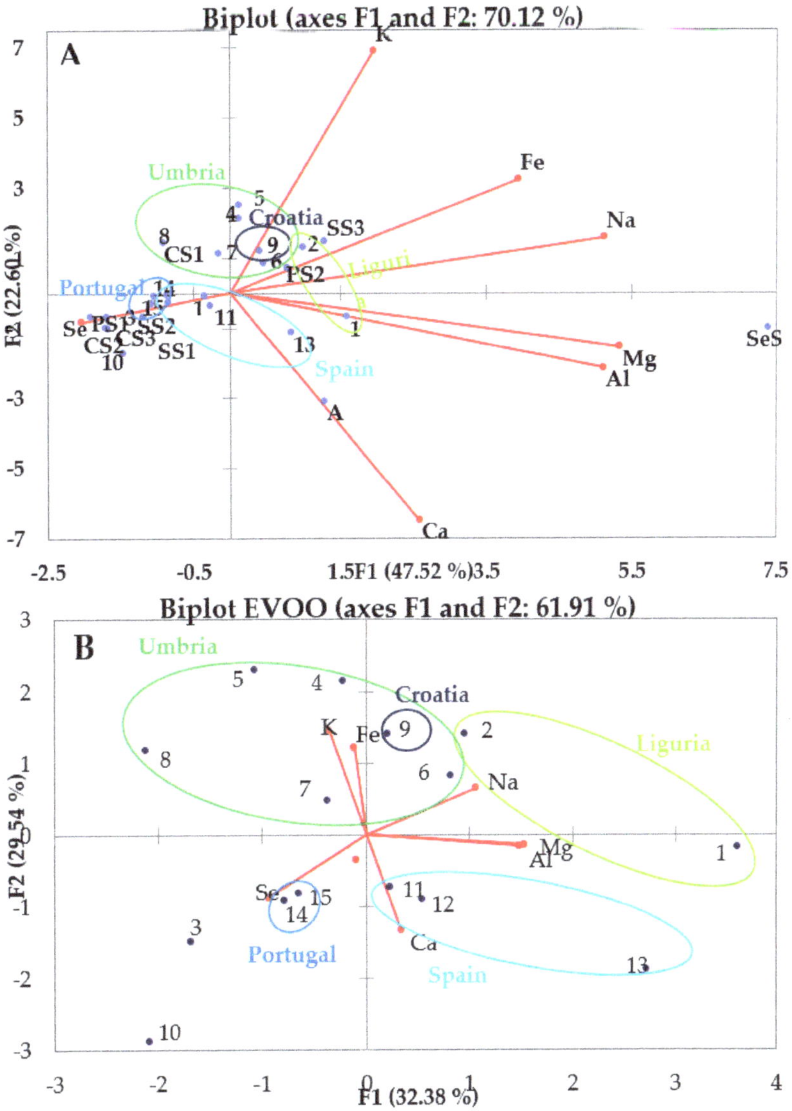

Figure 6. (**A**) PCA of inorganic component in all the investigated oils; (**B**) PCA of inorganic component in EVOOs. EVOO1, 2: Liguria, EVOO 3-8: Umbria, EVOO9: Croatian, EVOO10, 11, 12, 13: Spain, EVOO14, 15: Portugal. VOs–SS1: Sunflower Seed Oil (Brand A), SS2: Sunflower Seed Oil (Brand B), SS3: Sunflower Seed Oil (Brand C), CS1: Corn Seed Oil (Brand A), CS2: Corn Seed Oil (Brand B), CS3: Corn Seed Oil (Brand D), PS1: Peanut Seed Oil (Brand A), PS2: Peanut Seed Oil (Brand E), A: Almond Oil (Brand F), SeS: Sesame Seed Oil (Brand G).

3.3.2. PCA of CV and SWV Results

As regards the processing of data obtained by voltammetry measurements, a matrix with the potential values and the corresponding current values obtained experimentally for each sample was constructed. The variables are represented by the potential values, and the scores are obtained as linear combinations of the current values at different potentials.

The chemometric treatment was applied to the results obtained both by CV (Figure 7) and by SWV (Figures 8 and 9).

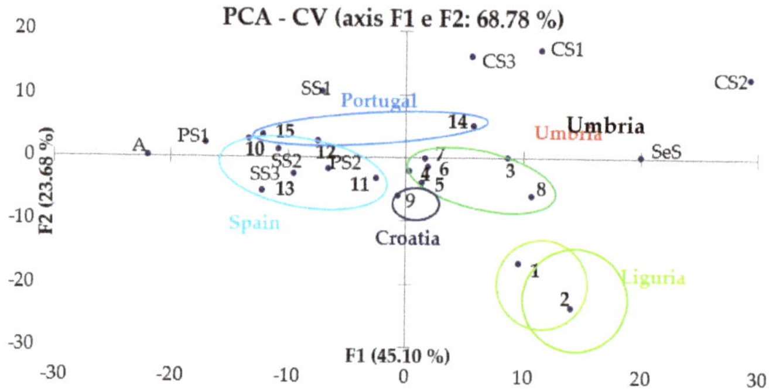

Figure 7. PCA of CV data.

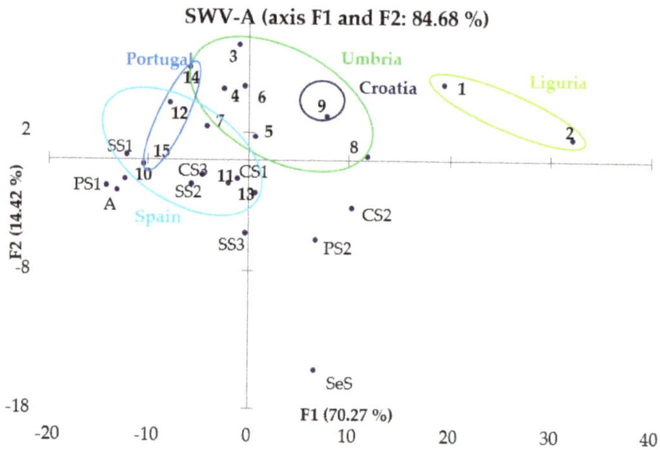

Figure 8. PCA of SWV-A data.

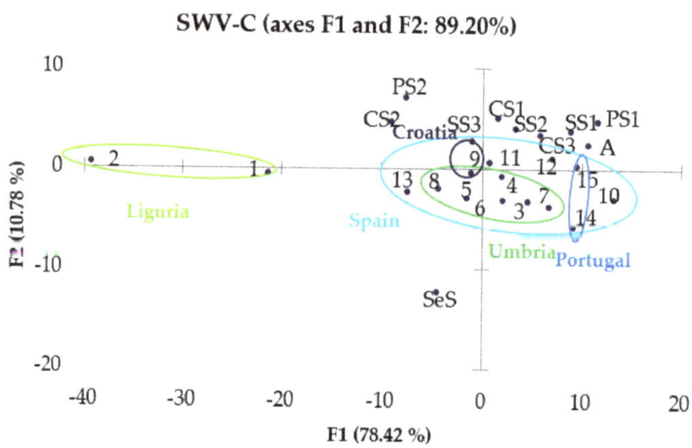

Figure 9. PCA of SWV-C data.

From the observation of these graphs, it is possible to note the presence of a good separation between the samples of EVOO and those of other oils that was expected from the remarks made in Section 3.2: in fact, it seems that the EVOOs occupy a more defined region of the vector space, while the oils of other nature are positioned in more diffused way around the EVOOs. In PC graphs obtained from the results of SWV-A, SWV-C and CV, Ligurian oils (circled in light green) tend to distance themselves from the other EVOOs, but always occupy a region of the vector space opposite to the oils of different botanical origin. In conclusion, there is the possibility of successfully discriminating not only EVOOs from oils of other natures, but also Ligurian EVOOs from Umbrian ones instead, whereas it is not so easy to distinguish Portuguese from Spanish EVOOs.

The method can therefore be useful for a botanical and merceological characterization.

4. Conclusions

The first part of this work was aimed at characterizing EVOOs samples from the point of view of the inorganic profile. The content of fourteen elements, namely Al, Ba, Ca, Cu, Fe, K, Li, Mn, Mg, Na, P, Sb, Se and Zn was quantified by ICP-OES

Ca, Mg, Na and Se are the major elements (>1 mg kg^{-1}). K, on the other hand, had a very variable trend. The data show that Italian EVOOs have greater Fe content than non-Italian ones, and Umbrian EVOOs show higher K but lower Na Al concentrations than Ligurian ones. The composition of the oils obtained from other plants is rather heterogeneous and no common trend can be identified.

The chemometric treatment of the experimental results via a PCA confirmed the possibilities offered by the method with regard to the discrimination of geographical origin on the basis of the proximity of the different scores, although some samples failed to meet expectations because of a higher concentration of Se. In particular, the behaviour of the two samples from Portugal shows the possibility to distinguish oils provided by different producers if the soil and the other characteristics of the area cultivated by each producer differ from each other. In synthesis, a distinction at local scale is feasible.

A geographic and varietal characterization is therefore possible even if the analysis is critical because of the complex organic matrix. Thus, the elementary analysis of oils can be a means to evaluate product authenticity, for example by detecting cuts with oils of different botanical species, or to determine the pollution from heavy metals derived from the soil in which olive trees were grown or from accidental contamination during the extraction or distribution phases. Hence, it can be a useful tool for monitoring the quality (from the point of view of contamination by metals) and the genuineness (understood as absence of sophistication) of the oil itself.

The presence of species with antioxidant properties, such as polyphenols and tocopherols, was detected from the peaks displayed in the voltammograms. SW provides more intense peaks than CV. The peaks in the profiles of different oils are different in shape, position, or intensity, because they are associated with antioxidant molecules of a different nature (even if they share common features, i.e., the range of potential in which they appear) and are present in different concentrations, which in turn are associated with different compositions of the species from which the oil derives. Therefore, voltammetry allowed us to discriminate different types of oils according to their botanical origin, both through the simple observation of the voltammetric profiles, and by the chemometric treatment of the results by PCA.

Chemometric treatment with PCA confirmed that the different oils are separated quite well in the score plot according to their botanical origin and in some cases also for their geographic origin. In fact, not only the Spanish EVOOs were distinguished from the Italian ones, but the Umbrian samples were also differentiated from the Ligurian ones, while the separation between Spanish and Portuguese EVOO was not as efficient.

As expected, non-EVOO samples were differentiated according to the plant species of origin.

In turn, these differences in chemical composition affect both the organoleptic characteristics and the stability of the oils. The antioxidant molecules, in fact, can take part in oxidation reactions to which the vegetable oils are naturally subject, protecting them from lipoperoxidation but also giving a bitter and pungent taste to the oil. A high content in metals involves a catalysis of the lipoperoxidation reactions that cause the early rancidity of the oil.

In this sense, voltammetry can be used as a monitoring tool for the production and distribution chain of oil and derivatives, assessing their conservation status, as well as for the identification of possible sophistication with oils of different botanical origins that would vary the intensity and the position of the peaks proper to the EVOO.

In conclusion, voltammetry presents good potential from the standpoint of monitoring the quality, state of conservation and the organoleptic properties of oils and their by-products. Possible future directions resulting from study includes, for the inorganic profile, the analysis of both oils and the soil in which the olives were grown; and for the redox profile, the analysis of a higher number of samples to better investigate the potential of the technique.

Supplementary Materials: The following supporting information can be downloaded at: https://www.mdpi.com/article/10.3390/molecules28010153/s1, Figure S1: SWV-A of Liguria oils; Figure S2: SWV-A of Umbria oils; Figure S3: SWV-A of EVOOs from Portugal and Croatia; Figure S4: SWV-A of Spain oils; Figure S5: SWV-C of Liguria oils; Figure S6: SWV-C of Umbria oils; Figure S7: SWV-C of Portugal and Croatia oils; Figure S8: CV of Spain oils; Figure S9: CV of Liguria oils; Figure S10: CV of Umbria oils; Figure S11: CV of Portugal and Croatia oils; Figure S12: CV of Spain oils; Figure S13: SWV-A of sunflower oils; Figure S14: SWV-A of corn seed oils; Figure S15: SWV-A of peanut seed oils; Figure S16: SWV-A of almond and sesame seed oils; Figure S17: SWV-C of sunflower oils; Figure S18: SWV-C of corn seed oils; Figure S19: SWV-C of peanut oils; Figure S20: SWV-C of almond and sesame seed oils; Figure S21: CV of sunflower seed oil; Figure S22: CV of corn seed oils; Figure S23: CV of peanut seed oils; Figure S24: CV of almond and sesame seed oils.

Author Contributions: A.G.: Methodology, data curation, writing—original draft, project administration. P.I.: Conceptualization, writing—original draft, writing—review and editing. G.S.: Investigation. A.D.: writing—review and editing. S.B.: Data Curation. E.G.: Writing—review and editing. M.M.: Data curation, writing—review and editing. F.S.: Investigation. O.A.: methodology writing—original draft, writing—review and editing. All authors have read and agreed to the published version of the manuscript.

Funding: This research was funded by AGER foundation, GRANT: 2016-0169, VIOLIN project: Valorization of Italian Olive products through Innovative analytical tools.

Institutional Review Board Statement: Not applicable.

Informed Consent Statement: Not applicable.

Data Availability Statement: Data, associated metadata, and calculation tools are available from the corresponding author (paolo.inaudi@unito.it).

Conflicts of Interest: The authors declare that they have no conflicts of interest.

References

1. FAO. *WHO Codex Commitee on Food Import and Export Inspection and Certification Systems*; Food and Agriculture Organization: Rome, Italy, 2018.
2. Dugo, G.; Lapera, L.; Latorre, G.; Giuffrida, D. Determination of Cd(II), Cu(II), Pb(II), and Zn(II) Content in Commercial Vegetable Oils Using Derivative Potentiometric Stripping Analysis. *Food Chem.* **2004**, *87*, 639–645. [CrossRef]
3. Charlebois, S.; Schwab, A.; Henn, R.; Huck, C.W. Food Fraud: An Exploratory Study for Measuring Consumer Perception towards Mislabeled Food Products and Influence on Self-Authentication Intentions. *Trends Food Sci. Technol.* **2016**, *50*, 211–218. [CrossRef]
4. Portarena, S.; Leonardi, L.; Scartazza, A.; Lauteri, M.; Baldacchini, C.; Farinelli, D.; Famiani, F.; Ciolfi, M.; Brugnoli, E. Combining Analysis of Fatty Acid Composition and Δ13C in Extra-Virgin Olive Oils as Affected by Harvest Period and Cultivar: Possible Use in Traceability Studies. *Food Control* **2019**, *105*, 151–158. [CrossRef]
5. Aued-Pimentel, S.; Takemoto, E.; Antoniassi, R.; Gastaldo Badolato, E.S. Composition of Tocopherols in Sesame Seed Oil: An Indicative of Adulteration. *Grasas Aceites* **2006**, *57*, 205–210. [CrossRef]

6. Aparicio-Ruiz, R.; García-González, D.L.; Lobo-Prieto, A.; Aparicio, R. Andalusian Protected Designations of Origin of Virgin Olive Oil: The Role of Chemical Composition in Their Authentication. *Eur. J. Lipid Sci. Technol.* **2019**, *121*, 1800133. [CrossRef]
7. European Commission. Available online: https://ec.europa.eu/info/index_en (accessed on 15 November 2022).
8. Di Serio, M.G.; Di Giacinto, L.; Di Loreto, G.; Giansante, L.; Pellegrino, M.; Vito, R.; Perri, E. Chemical and Sensory Characteristics of Italian Virgin Olive Oils from Grossa Di Gerace Cv: Grossa Di Gerace' Characterization. *Eur. J. Lipid Sci. Technol.* **2016**, *118*, 288–298. [CrossRef]
9. Tura, D.; Failla, O.; Bassi, D.; Pedò, S.; Serraiocco, A. Environmental and Seasonal Influence on Virgin Olive (Olea Europaea L.) Oil Volatiles in Northern Italy. *Sci. Hortic.* **2009**, *122*, 385–392. [CrossRef]
10. Morelló, J.-R.; Romero, M.-P.; Motilva, M.-J. Effect of the Maturation Process of the Olive Fruit on the Phenolic Fraction of Drupes and Oils from Arbequina, Farga, and Morrut Cultivars. *J. Agric. Food Chem.* **2004**, *52*, 6002–6009. [CrossRef]
11. Morelló, J.-R.; Motilva, M.-J.; Tovar, M.-J.; Romero, M.-P. Changes in Commercial Virgin Olive Oil (Cv Arbequina) during Storage, with Special Emphasis on the Phenolic Fraction. *Food Chem.* **2004**, *85*, 357–364. [CrossRef]
12. Lercker, G.; Bendini, A.; Cerretani, L. Quality, Composition and Production Process of Virgin Olive Oils. *Prog. Nutr.* **2007**, *9*, 134–148.
13. Fasciotti, M.; Pereira Netto, A.D. Optimization and Application of Methods of Triacylglycerol Evaluation for Characterization of Olive Oil Adulteration by Soybean Oil with HPLC–APCI-MS–MS. *Talanta* **2010**, *81*, 1116–1125. [CrossRef] [PubMed]
14. Dais, P.; Hatzakis, E. Quality Assessment and Authentication of Virgin Olive Oil by NMR Spectroscopy: A Critical Review. *Anal. Chim. Acta* **2013**, *765*, 1–27. [CrossRef] [PubMed]
15. Gromadzka, J.; Wardencki, W. Trends in edible vegetable oils analysis part b application of different analytical techniques. *Pol. J. Food Nutr. Sci.* **2011**, *61*, 89–99. [CrossRef]
16. Yildiz Tiryaki, G.; Ayvaz, H. Quantification of Soybean Oil Adulteration in Extra Virgin Olive Oil Using Portable Raman Spectroscopy. *J. Food Meas. Charact.* **2017**, *11*, 523–529. [CrossRef]
17. Nenadis, N.; Tsimidou, M.Z. Perspective of Vibrational Spectroscopy Analytical Methods in On-Field/Official Control of Olives and Virgin Olive Oil: Vibrational Spectroscopy in Virgin Olive Oil Analysis. *Eur. J. Lipid Sci. Technol.* **2017**, *119*, 1600148. [CrossRef]
18. Sánchez-López, E.; Sánchez-Rodríguez, M.I.; Marinas, A.; Marinas, J.M.; Urbano, F.J.; Caridad, J.M.; Moalem, M. Chemometric Study of Andalusian Extra Virgin Olive Oils Raman Spectra: Qualitative and Quantitative Information. *Talanta* **2016**, *156–157*, 180–190. [CrossRef]
19. Guzmán, E.; Baeten, V.; Pierna, J.A.F.; García-Mesa, J.A. Determination of the Olive Maturity Index of Intact Fruits Using Image Analysis. *J. Food Sci. Technol.* **2015**, *52*, 1462–1470. [CrossRef]
20. Gu, H.; Sun, Y.; Liu, S.; Li, S.; Zhang, W. A Feasibility Study of the Rapid Evaluation of Oil Oxidation Using Synchronous Fluorescence Spectroscopy. *Food Anal. Methods* **2018**, *11*, 3464–3470. [CrossRef]
21. Lia, F.; Castellano, A.M.; Zammit-Mangion, M.; Farrugia, C. Application of Fluorescence Spectroscopy and Chemometric Models for the Detection of Vegetable Oil Adulterants in Maltese Virgin Olive Oils. *J. Food Sci. Technol.* **2018**, *55*, 2143–2151. [CrossRef]
22. Rohman, A.; Man, Y.B.C. Application of Fourier Transform Infrared Spectroscopy for Authentication of Functional Food Oils. *Appl. Spectrosc. Rev.* **2012**, *47*, 1–13. [CrossRef]
23. Vasconcelos, M.; Coelho, L.; Barros, A.; de Almeida, J.M.M.M. Study of Adulteration of Extra Virgin Olive Oil with Peanut Oil Using FTIR Spectroscopy and Chemometrics. *Cogent Food Agric.* **2015**, *1*, 1018695. [CrossRef]
24. Torrecilla, J.S.; Rojo, E.; Domínguez, J.C.; Rodríguez, F. A Novel Method to Quantify the Adulteration of Extra Virgin Olive Oil with Low-Grade Olive Oils by UV−Vis. *J. Agric. Food Chem.* **2010**, *58*, 1679–1684. [CrossRef] [PubMed]
25. Jiang, L.; Zheng, H.; Lu, H. Application of UV Spectrometry and Chemometric Models for Detecting Olive Oil-Vegetable Oil Blends Adulteration. *J. Food Sci. Technol.* **2015**, *52*, 479–485. [CrossRef]
26. Zhu, F.; Fan, W.; Wang, X.; Qu, L.; Yao, S. Health Risk Assessment of Eight Heavy Metals in Nine Varieties of Edible Vegetable Oils Consumed in China. *Food Chem. Toxicol.* **2011**, *49*, 3081–3085. [CrossRef] [PubMed]
27. Zeiner, M.; Steffan, I.; Cindric, I.J. Determination of Trace Elements in Olive Oil by ICP-AES and ETA-AAS: A Pilot Study on the Geographical Characterization. *Microchem. J.* **2005**, *81*, 171–176. [CrossRef]
28. Pereira, J.S.F.; Moraes, D.P.; Antes, F.G.; Diehl, L.O.; Santos, M.F.P.; Guimarães, R.C.L.; Fonseca, T.C.O.; Dressler, V.L.; Flores, É.M.M. Determination of Metals and Metalloids in Light and Heavy Crude Oil by ICP-MS after Digestion by Microwave-Induced Combustion. *Microchem. J.* **2010**, *96*, 4–11. [CrossRef]
29. Zeiner, M.; Juranovic-Cindric, I.; Škevin, D. Characterization of Extra Virgin Olive Oils Derived from the Croatian Cultivar Oblica. *Eur. J. Lipid Sci. Technol.* **2010**, *112*, 1248–1252. [CrossRef]
30. Llorent-Martínez, E.J.; Ortega-Barrales, P.; Fernández-de Córdova, M.L.; Domínguez-Vidal, A.; Ruiz-Medina, A. Investigation by ICP-MS of Trace Element Levels in Vegetable Edible Oils Produced in Spain. *Food Chem.* **2011**, *127*, 1257–1262. [CrossRef]
31. Bajoub, A.; Medina-Rodríguez, S.; Gómez-Romero, M.; Ajal, E.A.; Bagur-González, M.G.; Fernández-Gutiérrez, A.; Carrasco-Pancorbo, A. Assessing the Varietal Origin of Extra-Virgin Olive Oil Using Liquid Chromatography Fingerprints of Phenolic Compound, Data Fusion and Chemometrics. *Food Chem.* **2017**, *215*, 245–255. [CrossRef]
32. Frankel, E.N. Chemistry of Extra Virgin Olive Oil: Adulteration, Oxidative Stability, and Antioxidants. *J. Agric. Food Chem.* **2010**, *58*, 5991–6006. [CrossRef]

33. Jabeur, H.; Drira, M.; Rebai, A.; Bouaziz, M. Putative Markers of Adulteration of Higher-Grade Olive Oil with Less Expensive Pomace Olive Oil Identified by Gas Chromatography Combined with Chemometrics. *J. Agric. Food Chem.* **2017**, *65*, 5375–5383. [CrossRef] [PubMed]
34. Salghi, R.; Armbruster, W.; Schwack, W. Detection of Argan Oil Adulteration with Vegetable Oils by High-Performance Liquid Chromatography–Evaporative Light Scattering Detection. *Food Chem.* **2014**, *153*, 387–392. [CrossRef] [PubMed]
35. Melucci, D.; Bendini, A.; Tesini, F.; Barbieri, S.; Zappi, A.; Vichi, S.; Conte, L.; Toschi, T.G. Rapid Direct Analysis to Discriminate Geographic Origin of Extra Virgin Olive Oils by Flash Gas Chromatography Electronic Nose and Chemometrics. *Food Chem.* **2016**, *204*, 263–273. [CrossRef]
36. López Ortíz, C.M.; Prats Moya, M.S.; Berenguer Navarro, V. A Rapid Chromatographic Method for Simultaneous Determination of β-Sitosterol and Tocopherol Homologues in Vegetable Oils. *J. Food Compos. Anal.* **2006**, *19*, 141–149. [CrossRef]
37. Stilo, F.; Liberto, E.; Reichenbach, S.E.; Tao, Q.; Bicchi, C.; Cordero, C. Untargeted and Targeted Fingerprinting of Extra Virgin Olive Oil Volatiles by Comprehensive Two-Dimensional Gas Chromatography with Mass Spectrometry: Challenges in Long-Term Studies. *J. Agric. Food Chem.* **2019**, *67*, 5289–5302. [CrossRef]
38. Valli, E.; Bendini, A.; Berardinelli, A.; Ragni, L.; Riccò, B.; Grossi, M.; Gallina Toschi, T. Rapid and Innovative Instrumental Approaches for Quality and Authenticity of Olive Oils: Innovative Approaches for Quality of Virgin Olive Oils. *Eur. J. Lipid Sci. Technol.* **2016**, *118*, 1601–1619. [CrossRef]
39. Zappi, D.; Gabriele, S.; Gontrani, L.; Dini, D.; Sadun, C.; Marini, F.; Antonelli, M.L. Biologically Friendly Room Temperature Ionic Liquids and Nanomaterials for the Development of Innovative Enzymatic Biosensors: Part II. *Talanta* **2019**, *194*, 26–31. [CrossRef]
40. Zappi, D.; Sadun, C.; Gontrani, L.; Dini, D.; Antonelli, M.L. A New Electrochemical Sensor for Extra-Virgin Olive Oils Classification. *Food Control* **2020**, *109*, 106903. [CrossRef]
41. Jamali, M.K.; Kazi, T.G.; Arain, M.B.; Afridi, H.I.; Jalbani, N.; Sarfraz, R.A.; Baig, J.A. A Multivariate Study: Variation in Uptake of Trace and Toxic Elements by Various Varieties of Sorghum Bicolor L. *J. Hazard. Mater.* **2008**, *158*, 644–651. [CrossRef]
42. Mourhat, Z.; Touzara, S.; Maallah, R.; Mbarki, M.; Chtaini, A. Electrochemical Evaluation of the Antioxidant Capacity of Phenolic Compounds in Virgin Olive Oil. *J. Biosens. Bioelectron.* **2017**, *8*, 2. [CrossRef]
43. Cimato, A.; Dello Monaco, D.; Distante, C.; Epifani, M.; Siciliano, P.; Taurino, A.M.; Zuppa, M.; Sani, G. Analysis of Single-Cultivar Extra Virgin Olive Oils by Means of an Electronic Nose and HS-SPME/GC/MS Methods. *Sens. Actuators B Chem.* **2006**, *114*, 674–680. [CrossRef]
44. Inaudi, P.; Giacomino, A.; Malandrino, M.; La Gioia, C.; Conca, E.; Karak, T.; Abollino, O. The Inorganic Component as a Possible Marker for Quality and for Authentication of the Hazelnut's Origin. *Int. J. Environ. Res. Public Health* **2020**, *17*, 447. [CrossRef] [PubMed]
45. Gaggero, E.; Malandrino, M.; Fabbri, D.; Bordiglia, G.; Fusconi, A.; Mucciarelli, M.; Inaudi, P.; Calza, P. Uptake of Potentially Toxic Elements by Four Plant Species Suitable for Phytoremediation of Turin Urban Soils. *Appl. Sci.* **2020**, *10*, 3948. [CrossRef]
46. Barabasz, W.; Albi, D.; Jaśkowska, M.; Lipiec, J. Ecotoxicology of Aluminium. *Pol. J. Environ. Stud.* **2002**, *11*, 199–203.
47. Rodríguez-Méndez, M.L.; Apetrei, C.; de Saja, J.A. Evaluation of the Polyphenolic Content of Extra Virgin Olive Oils Using an Array of Voltammetric Sensors. *Electrochim. Acta* **2008**, *53*, 5867–5872. [CrossRef]

molecules MDPI

Article

Olive Oil Benefits from Sesame Oil Blending While Extra Virgin Olive Oil Resists Oxidation during Deep Frying

Evangelia T. Ioannou [1], Konstantinos S. Gliatis [1], Evangelos Zoidis [2] and Constantinos A. Georgiou [1,3,*]

1 Chemistry Laboratory, Department of Food Science and Human Nutrition, Agricultural University of Athens, 75 Iera Odos, 118 55 Athens, Greece; evioannou@aua.gr (E.T.I.)
2 Laboratory of Nutritional Physiology and Feeding, Department of Animal Science, Agricultural University of Athens, 75 Iera Odos, 118 55 Athens, Greece; ezoidis@aua.gr
3 FoodOmics.GR Research Infrastructure, 118 55 Athens, Greece
* Correspondence: cag@aua.gr

Abstract: Fresh potatoes were deep-fried in olive oil (OO), extra virgin olive oil (EVOO), and their blends with 5%, 10%, and 20% v/v sesame oil (SO). This is the first report on the use of sesame oil as a natural source of antioxidants during olive oil deep frying. The oil was evaluated for anisidine value (AV), free fatty acids (FFAs), extinction coefficient (K_{232} and K_{270}), Trolox equivalent antioxidant capacity (TEAC), and total phenols (TPs) until the total polar compounds (TPCs) reached 25%. Sesame lignan transformations were monitored through reversed-phase HPLC. While the TPCs in olive oils increased at a steady rate, the addition of 5%, 10%, and 20% v/v SO delayed TPCs' formation for 1, 2, and 3 h, respectively. The addition of 5%, 10%, and 20% v/v SO increased the olive oil frying time by 1.5 h, 3.5 h, and 2.5 h, respectively. The addition of SO to OO reduced the secondary oxidation products' formation rate. The AV for EVOO was lower than OO and all tested blends, even those with EVOO. EVOO was more resistant to oxidation than OO, as measured by the TPCs and TEAC, while the frying time rose from 21.5 to 25.25 h when EVOO replaced OO. The increase in frying time for OO but not for EVOO, after SO addition, points to a niche market for EVOO in deep frying.

Keywords: antioxidants; deep frying; olive oil; anisidine value; sesame oil; sesame lignans; total polar compounds

Citation: Ioannou, E.T.; Gliatis, K.S.; Zoidis, E.; Georgiou, C.A. Olive Oil Benefits from Sesame Oil Blending While Extra Virgin Olive Oil Resists Oxidation during Deep Frying. *Molecules* **2023**, *28*, 4290. https://doi.org/10.3390/molecules28114290

Academic Editor: Gavino Sanna

Received: 27 April 2023
Revised: 15 May 2023
Accepted: 19 May 2023
Published: 24 May 2023

1. Introduction

Deep frying is one of the world's most popular culinary processes, both for industrial and domestic food preparation purposes. During deep frying, food is immersed in hot oil at temperatures of 150 to 190 °C. In the presence of air, many complex reactions take place such as oxidation, hydrolysis, and polymerization [1–3]. These reactions influence the quality of the final product, such as the flavor, texture, shelf life of the oil, and nutrient composition, with potential adverse effects on human health [4]. The type of frying oil, its chemical composition, and its physical and physicochemical properties are major parameters that influence the chemical reactions and determine the performance of the frying oil against oxidation and decomposition [5].

Over the ages, olive oil has been widely produced and consumed in Mediterranean countries, with it being the main lipid source in the Mediterranean diet. Its beneficial properties are associated with fatty acid composition, phenolic antioxidants, and other minor compounds that make olive oil a very interesting option among oils and fats [6,7]. Extra virgin olive oil exhibits high resistance to oxidation in comparison with other vegetable oils, and it is well known for its very good sensory and health properties [8,9]. Olive oil is resistant to degradation under domestic frying conditions, independently of its category label [10]. Olive oil's naturally occurring antioxidants play a significant role in the thermal stability during deep frying [11]. Synthetic and natural antioxidants can be added to prevent or minimize the oxidative deterioration of the oil. The most commonly used antioxidants are butylated hydroxy anisole (BHA), butylated hydroxytoluene (BHT), propyl gallate (PG), and tert-butylhydroquinone (TBHQ). However, the use of synthetic antioxidant additives is regulated in most countries because of concerns regarding their long-term health effects. Natural components in foods with radical scavenging or antioxidant activity have attracted interest as alternatives to synthetic antioxidants. The addition of sources of natural antioxidants can possibly be used to improve olive oil's resistance to the formation of primary and secondary oxidation products without making considerable changes to their natural composition. Blending different types of vegetable oils may help to extend the thermal stability and nutritional profile of frying oils [12]. Sesame oil demonstrates higher oxidative stability than other vegetable oils [2,13]. The study of this superior oxidative stability has mainly been focused on sesame lignans, which are present in small amounts in sesame oil. Sesamin and sesamolin are the major lignans found in sesame seeds [14,15]. When sesame seeds are roasted at a high temperature, sesamolin degrades into sesamol [16]. Sesamol is reported to possess higher radical scavenging activity compared to sesamin and sesamolin [17]. The significant stability of sesame oil could be related with the continuous generation of sesamol from the degradation of sesamolin during thermal oxidation, rather than the initial antioxidant content [18].

In one study, the addition of sesame lignans in oils during frying increased sesamol and decreased sesamolin, while sesamin was rather resistant to heat. Sesame lignans could have applications as natural antioxidants in the edible oil and food industry [19]. Other studies suggest that lignan compounds in sesame oil are effective antioxidants in deep fat frying due to their high stability and efficacy [2,20]. The addition of roasted sesame oil, as a natural source of antioxidants, prolonged the oil's heat stability and shelf life. Moreover, when the roasted sesame oil concentration increased, the antioxidant capacity of frying oils increased. The proper blending of high polyunsaturated sunflower oil with sesame oil can produce oil blends of high nutritional value and enhanced stability for home cooking and deep frying [21].

Consumers' perception of the health benefits of olive oil creates a niche market for olive oil for deep frying, despite its high price. Consumers nowadays may have a different perception of extra virgin olive oil, recognizing its superiority over olive oil. This creates the need to further study these two olive oil products during deep frying and assess the benefits of adding antioxidants as blends with natural products that will not deteriorate consumers' perception of the health benefits of olive oil. Although olive oil's deep-frying characteristics have been studied, we were not able to find any study with sesame oil admixtures. In this study, we describe the effect of adding virgin sesame oil as a source of natural antioxidants at different levels to extra virgin olive oil and olive oil during domestic deep frying. Sesame lignan changes in the oil samples during deep frying and the correlation between the sesame oil antioxidant activity and its efficacy in retarding the oxidative deterioration of olive oil and extra virgin olive oil were investigated. In order to evaluate the oxidation progress, we monitored the total polar compounds, anisidine value, free fatty acids, extinction coefficient, total phenols, and antioxidant activity.

2. Results and Discussion

2.1. Total Polar Compounds

The formation of polar compounds indicates oil deterioration and is strongly related to primary and secondary oxidation during frying. Total polar compounds are considered one of the most objective indicators in evaluating the deterioration of deep-frying oils [22,23]. In several countries, the rejection value for TPCs has been set at 25% by weight for frying oils [24]. TPCs increased linearly, as shown in Figure 1 for OO and Figure 2 for EVOO. The OO frying time increased from 21.5 to 23, 25, and 24 h upon the addition of 5, 10, and 20% SO, respectively. During this procedure, the formation of TPCs was slightly delayed when sesame oil was added. The delay was proportional to the SO concentration, lasting 1, 2, and 3 h for 5, 10, and 20% SO, respectively. Then, TPCs increased at a steady rate of 1% per hour for all three blends, which is lower in comparison to 1.5% for pure OO.

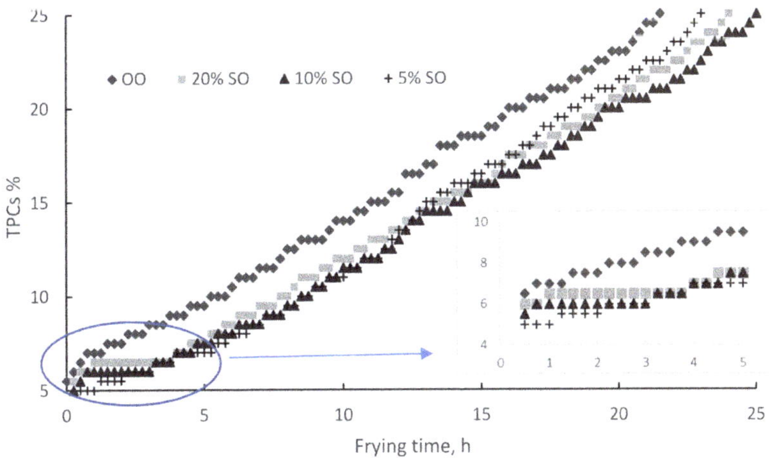

Figure 1. Total polar compounds during deep frying: olive oil and sesame oil blends.

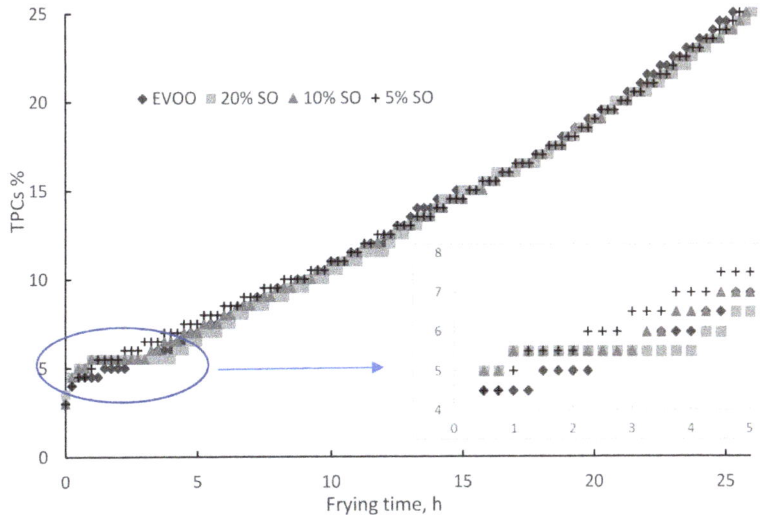

Figure 2. Total polar compounds during deep frying: extra virgin olive oil and blends with sesame oil.

The EVOO frying time was not increased by SO blending and was approximately the same (25–26 h) for all blends. This is due to the higher antioxidant content of EVOO, indicating the higher nutritional value and health benefits of EVOO in comparison to OO. The TPC formation rate for EVOO did not decrease upon the addition of SO; it was always 1% per hour. A slight delay in the formation of TPCs upon 10 and 20% SO addition to EVOO was also detected, while 5% had no effect. After this delay, TPCs increased at a steady rate of 1% per hour (Figure 2).

2.2. *Lignans*

Large amounts of sesamol are produced from sesamolin during frying, contributing to oil stability [16,25]. Chatzos and Georgiou [26] reported that radical scavenging activity increases during virgin sesame oil heating, in contrast to all other seed oils in their study. This paradox is attributed to the fact that sesamol has higher antioxidant activity than its precursor, sesamolin [17]. We followed the decomposition of sesamolin to sesamol during the domestic deep frying of potatoes in admixtures with olive oil through reverse-phase HPLC. Initially, only sesamin and sesamolin were detected (Figure 3). After 1h of frying, sesamol was detected and sesamolin decreased by 18%. After 2 h of deep frying, both sesamol and sesamolin decreased by 5% and 38%, respectively. After 4 h, sesamol and sesamolin decreased further by 60% and 70%, respectively. A similar study on sesame oil by Hemalatha and Ghafoorunissa is in accordance with our study, reporting the maximum concentration of sesamol after one hour followed by a gradual decrease for one more hour [27].

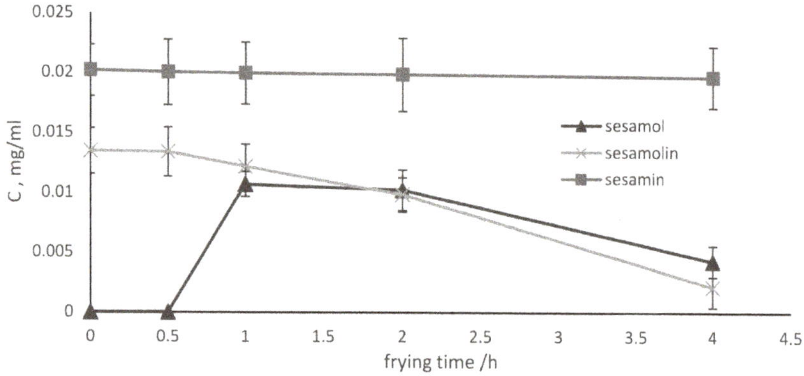

Figure 3. Sesame lignans during deep frying using 20% v/v sesame oil in olive oil.

As mentioned in Section 3.1, the TPCs after one hour of frying remained constant at the same time as when the sesamol concentration peaked. Then, after four hours, when sesamol was depleted, the TPCs increased again, while all blends exhibited similar behavior. This points to a beneficial role of sesame oil addition. The delay of TPC formation correlates to sesamolin concentrations: higher concentrations increase the delay time. This finding is in agreement with previous research, where the antioxidant capacity of sesamol may have been influenced by the concentration [28].

2.3. *Changes in Anisidine Value*

Deep frying promotes secondary oxidation products, mostly conjugated dienals and 2-alkenals, which are more stable during the heating process. Thus, the AV is an essential and reliable test to measure oil oxidation [29]. These compounds are accessed through the anisidine value. The AV increased significantly for both the OO and the blends with SO during the initial 12h. Then, the AV increased at a lower rate. It should be noted that the blends always scored lower than OO (Table 1).

The AV significantly increased for EVOO and all its SO blends during the initial 12 h (Table 1). At the same time, the blends scored higher in the AV. Then, the AV increased at a lower rate, while the blends still scored higher. It is interesting to note that secondary oxidation products were lower in EVOO throughout the study. This points out that the antioxidant compounds therein are much more efficient than sesame oil.

Table 1. Changes in AV during deep frying. Olive oil, extra virgin olive oil, and blends with sesame oil.

Time	Frying Cycle	Anisidine Value							
		OO	OO Blends with SO			EVOO	EVOO Blends with SO		
			5%	10%	20%		5%	10%	20%
0	1	26.3	18.9	12.2	23.5	16.0	20.4	20.1	36.6
0.5	3	30.7	21.9	14.3	24.0	19.7	30.9	28.7	40.7
1	5	45.8	26.7	15.3	24.3	21.2	36.6	34.9	42.0
1.5	7	52.1	29.6	19.4	33.4	23.6	44.7	38.5	41.7
2.5	11	54.0	38.1	26.9	36.1	29.1		45.7	45.6
4	17	55.3	48.8	37.9	46.0	41.4	54.2	45.9	56.1
4.5	19	52.5	48.9	41.5	48.7	47.2	54.6	46.4	59.65
9.5	39	95.9		63.0	76.7	71.8		66.9	74.9
12	49	97.8	85.7	69.5	80.6	64.3	67.6	68.5	90.9
14	57	102.0	90.0	75.9	90.3	76.3	75.2	88.4	97.1
16	65	109.2	94.3	79.4	96.4	77.8	86.7	90.3	107.2
18	73	113.4		84.5	109.3	92.2	96.9	97.9	102.3
21.25	86	128.2							
23	93		114.1	96.2	124.6	97.1	105.3	100.0	122.5
24	97		124.9	102.0	126.8	111.9	115.9	117.8	
24.75	100					114.7	119.6		
25	101			112.3				119.9	129.2
26	105							121.6	129.9

OO: olive oil, EVOO: extra virgin olive oil, SO: sesame oil.

2.4. Free Fatty Acids

Hydrolysis in fats and oils results in the formation of free fatty acids, mono- and di-glycerides, and glycerol, inducing oxidative degradation and contributing to shelf life reduction [2]. The free fatty acid content increases during frying [30].

The FFA content in the OO sample was 0.71%, which is within the legally accepted limit of 1%. After 12h of frying, the FFA content in OO and the three blends increased to 1.44, 1.52, 1.47, and 1.59, respectively, as shown in Table 2. The total frying time was 21.5, 23, 25, and 24 h for OO and the three blends. The end FFA values were 1.9, 2.3, 2.8, and 3.0%, respectively. Although the frying time increased due to SO addition, it was accompanied by a higher FFA content, which could have a negative organoleptic impact or even surpass the maximum value set in specific countries (2.5%) [24].

The FFA content in EVOO was 0.67%, which is within the legally accepted limit of 0.8% established by the Commission Delegated Regulation (EU 2015/1830). After 12 h of deep frying, the FFA content in EVOO and the three blends increased to 1.29, 1.30, 1.45, and 1.41%, respectively, as shown in Table 2. The total frying time was 25, 25.25, 25.5, and 26 h for EVOO and the three blends. The end FFA values were 2.2, 1.8, 2.4, and 2.4%, respectively. The frying time was accompanied by a higher FFA content without exceeding the maximum value set in specific countries (2.5%) [24].

2.5. Extinction Coefficients: K_{232} and K_{270}

When polyunsaturated fatty acids are oxidized, ultraviolet absorption increases. The changes in ultraviolet absorption at 232 nm are associated with the formation of conjugated dienes of polyunsaturated fatty acids, while changes at 270 nm are associated with the formation of conjugated trienes and carbonyl compounds. OO's and EVOO's values of K_{232} and K_{270} are shown in Tables 2 and 3, respectively. The values for OO, EVOO, and their blends with SO increased during frying. There were no evident differences between the types of oil or between the blends. The increase was not related to the amount of SO. Moreover, no difference between OO and EVOO was recorded.

Table 2. Olive oil and blends with sesame oil, changes in FFA, extinction coefficients, and TEAC value.

Frying Time (h)	FFA	K_{232}	K_{270}	TEAC (mmol/Kg)
OO				
0	0.71 ± 0.03	1.55	0.20	0.91 ± 0.01
12	1.44 ± 0.08	3.91	1.77	0.15 ± 0.02
21.5	1.91 ± 0.05	3.99	1.86	0.26 ± 0.02
5% SO				
0	0.69 ± 0.00	1.19	0.21	1.10 ± 0.01
12	1.52 ± 0.04	3.91	1.08	0.02 ± 0.02
23	2.34 ± 0.05	3.90	1.21	0.60 ± 0.02
10% SO				
0	0.58 ± 0.08	1.43	0.26	1.29 ± 0.01
12	1.47 ± 0.05	2.24	1.29	0.099 ± 0.01
25	2.81 ± 0.1	4.15	1.74	0.53 ± 0.03
20% SO				
0	0.41 ± 0.01	2.09	0.36	1.37 ± 0.01
12	1.59 ± 0.06	2.16	1.61	0.18 ± 0.02
24	3.03 ± 0.04	4.13	2.27	0.92 ± 0.03

OO: olive oil, SO: sesame oil.

Table 3. Extra virgin olive oil and blends with sesame oil, changes in FFA, extinction coefficients, and TEAC value.

Frying Time (h)	FFA	K_{232}	K_{270}	TEAC (mmol/Kg)
EVOO				
0	0.67 ± 0.01	1.22	0.13	1.91 ± 0.02
12	1.29 ± 0.05	3.63	0.89	0.32 ± 0.01
25	2.21 ± 0.07	2.94	0.79	0.49 ± 0.03
5% SO				
0 h	0.67 ± 0.00	2.13	0.25	1.63 ± 0.01
12 h	1.30 ± 0.08	3.39	1.21	0.34 ± 0.02
25.25 h	1.79 ± 0.05	3.47	1.27	0.44 ± 0.02
10% SO				
0 h	0.63 ± 0.03	2.06	0.19	2.1 ± 0.02
12 h	1.45 ± 0.04	3.11	1.31	0.46 ± 0.02
25.5 h	2.44 ± 0.07	2.35	1.41	0.68 ± 0.03
20% SO				
0 h	0.63 ± 0.01	2.03	0.36	2.03 ± 0.01
12 h	1.41 ± 0.12	2.77	1.55	0.56 ± 0.02
26 h	2.42 ± 0.08	4.06	1.92	0.94 ± 0.02

EVOO: extra virgin olive oil, SO: sesame oil.

2.6. Changes in Antioxidant Capacity

Lipid oxidation is a free radical reaction that is strongly modulated by synthetic and natural antioxidant compounds. During deep frying, antioxidant compounds are consumed, resulting in a lower score in antioxidant capacity tests. Measured by the radical scavenging ability while reacting with a relatively stable radical such as 2,2-diphenyl-1-picrylhydrazyl (DPPH) [31].

In our study, the total antioxidant capacity was expressed as the Trolox equivalent antioxidant capacity (TEAC), defined as the mmol Trolox/kg of oil. The OO blends with SO initially had higher TEAC values than OO, which were proportional to the percentage of SO in the blend (Table 2). After 12 h of deep frying, all oils reached extremely low TEAC values. This comes in accordance with Kalantzakis et al.'s study, where olive oil samples practically lost their radical scavenging activity after 5 h of heating at 180 °C [32]. Although the oil samples were not expected to present any radical scavenging activity at the end of the frying experiment, the TEAC values were elevated for all samples (Figure S1a). This is an artifact explained by the reaction of DPPH with aldehydic compounds that are the end products of lipid oxidation and are produced in high concentrations during the late stage of deep frying [33].

The antioxidant capacity of EVOO was not increased that much with the addition of SO, as with OO (Tables 2 and 3, zero frying time). The antioxidant capacity of EVOO after 12 h of deep frying decreased by 84%. This improved to 72% upon 20% SO addition (Table 3). In a similar way as was said for OO above, the TEAC values increased after 25 h of deep frying (Figure S1b).

2.7. Changes in Total Phenols

Phenolics are the major health-promoting compounds in olive oil and the Mediterranean diet [34]. It is, therefore, important to follow the evolution of phenolics during deep frying to assess the potential health benefits of using olive oil.

The initial phenolic content of OO was 55.96 mg/Kg. The total phenols suffered a significant decrease for OO and all its blends with SO after 12 h of deep frying, going almost to zero (Figure S2).

The initial amount of phenolics in EVOO was 127.0 mg/Kg. After 12 h of deep frying, EVOO and all its blends with SO retained 30% of their phenolic content. After 25 h, there were no phenolic compounds left (Figure S3).

2.8. Fatty Acid Composition

The fatty acid compositions reflected the high proportion of oleic acid in OO (74.6%) and EVOO (75%). In contrast to olive oils, the major fatty acid in SO is linoleic (41.5%), followed by oleic (31.9%), palmitic (14.8%), and stearic (9%) acids. The fatty acid compositions of SO, OO, and EVOO were in agreement with values from the literature [35,36].

3. Materials and Methods

3.1. Reagents

Isooctane, isopropanol, glacial acetic acid, chloroform, ethyl acetate, acetone, diethyl ether, ethanol, n-hexane and methanol of analytical grade, sodium acetate of pro-analysis grade, and potassium hydroxide pellets for analysis were obtained from Merck, Darmstadt, Germany.

p-anisidine 99%, 2,2-Diphenyl-1-picrylhy-drazyl (DPPH) 90%, and 6-hydroxy-2,5,7,8-tetramethylchroman-2-car-boxylic acid (Trolox) 97% were purchased from Sigma-Aldrich (St. Louis, MO, USA). Gallic acid monohydrate was supplied from Riedel-de Haen (Seelze, Germany). LC-MS-grade methanol and CDCl3 98% D were purchased from Fluka. Sesamol was purchased from Sigma Chemical Co. (St. Louis, MO, USA).

3.2. Oil Samples

A commercial extra virgin olive oil (EVOO); a commercial blend of refined olive oil and virgin olive oil labeled as olive oil (OO) purchased from Minerva S.A. (Athens, Greece); and commercial virgin sesame oil (SO) purchased from Haitoglou Bros S.A. (Thessaloniki, Greece) were used for the frying experiments. The oils were purchased from local stores in Athens, Greece in sealed and marked commercial containers.

3.3. Frying Process

A 2.5 L domestic deep-frying electric fryer (KENWOOD DF520) was used, where the temperature was regulated at 170 ± 5 °C. Fresh potato strips (7 cm × 0.5 cm × 0.5 cm), of Spunta variety cultivated in Greece, were deep-fried in 70 g batches. The batches were fried at 9 min intervals for 12 h per day for two consecutive days, without oil replenishment, until the oil was discarded. The end-of-frying assay and oil rejection point were determined by the value of total polar compounds (max. 25%), according to the regulation of frying fats and oils in most European Countries [24]. The frying experiment was planned in such a way as to simulate a continuous/prolonged deep-frying process.

3.4. Oil Sampling

After each frying session, lasting 15 min, the TPCs were assessed directly on the hot oil. Extinction coefficients (K_{232} and K_{270}), antioxidant capacity (AC), free fatty acids (FFAs), and total phenols (TPs) were assessed by removing a 6 mL sample: (a) right after the thermal equilibration of the oil to 170 ± 5 °C before frying the potatoes, (b) at 12 h, and (c) at the time when the TPCs reached 25% and the oil was rejected. The anisidine value (AV) was assessed by removing 0.5 mL sample at regular intervals. The samples were placed in screw-cap glass vials and were immediately stored in the freezer until analysis. To monitor sesame lignan transformations, 2 mL oil samples were removed before starting and after 1 h, 2 h, and 4 h.

3.5. Total Polar Compounds

TPC estimation was based on the dielectric constant changes, measured directly on the hot oil with a Testo 270 sensor (Testo SE & Co. KGaA, Titisee-Neustadt, Germany). Before measuring, about 5 min was allowed after removing the fried potatoes until there were no more bubbles rising. The sensor took about 1 min to obtain a stable reading. TPC % content along with the temperature were displayed on the screen of the sensor. Sensor calibration was performed through the oil, supplied by the Testo 270 manufacturer, right before the analysis. The sensor was cleaned with warm water and neutral detergent and dried well between measurements.

3.6. HPLC Analysis

Sesamin and sesamolin were isolated and crystallized from sesame oil, as described by Hemalatha and Ghafoorunissa [26]. Sesamin and sesamolin were characterized with mass spectrometry and ^{1}H-NMR and ^{13}C-NMR spectrometry. Sesamol was purchased from Sigma Chemical Co. (St. Louis, MO, USA). HPLC analysis was carried out with an Agilent Technologies 1100 series diode array detector. Sesame lignans were analyzed as described by Wen-Huey Wu [37], using HPLC equipped with a Supelco Analytical HPLC column (Discovery HS C18) with a length of 250 mm, internal diameter of 4.6 mm, and particle size of 5 μm. The mobile phase was a mixture of methanol–deionized water (70:30, *v*/*v*) at a flow rate of 0.8 mL/min and the column temperature was maintained at 25 °C. The absorption at 290 nm was monitored. Twenty-microliter aliquots of oils, dissolved in chloroform (0.5 mg/mL), were injected for analysis. The retention times for standard sesamol, sesamin, and sesamolin were 4.6, 15.6, and 20.8 min, respectively (Figures S4–S9, Table S2).

3.7. Analytical Methods

The AV was determined according to the modified European Pharmacopoeia 5.0 method. The oil samples were dissolved in isooctane and then allowed to react for ten minutes with a 2.5 g/L p-anisidine solution in acetic acid. The absorbance value was measured at 350 nm using the Cary 60 Scan UV–visible spectrophotometer. The spectrophotometer gives the average absorbance of three readings.

The extinction coefficients (K_{232} and K_{270}) and free acidity of the oils were determined according to the analytical methods described in European Commission Regulations (Commission Regulation (EEC) No.2568/91) [38]. The total antioxidant capacity was determined through DPPH radical. A 65 mg oil sample was added to 4 mL of a 1.3×10^{-4} M solution of DPPH in ethyl acetate. Then, the mixture was shaken vigorously and left in darkness for 1 h. Finally, the absorbance of the mixture was measured against ethyl acetate (blank) at 515 nm with a spectrophotometer (Cary 60 Scan UV–visible spectrophotometer). The total antioxidant capacity was expressed as Trolox equivalent antioxidant capacity (TEAC), defined as the mmol Trolox/kg of oil [26]. The total phenols in the methanolic extract of the oils were determined colorimetrically at 765 nm with Folin–Ciocalteu reagent according to Capannesi et al. [39]. Gallic acid standard solutions were used for calibration (r = 0.9998).

3.8. Fatty Acid Analysis

The sample (0.1 g) was dissolved in 2 mL of heptane and shaken. The sample was then transesterified with 0.2 mL of 2 N methanolic potassium hydroxide solution and vigorously shaken (Commission Regulation (EEC) No.2568/91) [37]. Methyl esters in the upper layer were assessed with GC–FID. GC was performed on a 30 m × 0.25 mm i.d. × 0.25 μm film HP-5MS capillary column using a Hewlett Packard (Waldbronn, Germany) 5890 gas chromatograph connected to a flame ionization detector (FID). The GC conditions used were as follows: injection volume, 1 μL; split injection, 50:1 at 220 °C; and oven temperature ramped to 270 °C at 5 °C min^{-1} held for 5 min, giving a total run time of 55 min. The helium gas carrier was held at a constant flow rate of 1 mL min^{-1}, whilst the detector was set at a temperature of 290 °C.

4. Conclusions

Blending olive oil with sesame oil decreases secondary oxidation products throughout the frying period until its rejection at 25% total polar compounds. Through blending, the useful frying period increases from 21.5 h to 25 h in proportion to the amount of sesame oil, reaching the frying time of extra virgin olive oil, which does not benefit from blending.

Blending extra virgin olive oil, in contrast to olive oil, increases secondary oxidation products. It is clear that the natural antioxidant compounds in extra virgin olive oil are more efficient at preventing deep-frying oxidative damage than sesame oil blending.

In summary, sesame oil addition increases olive oil's frying time but is not beneficial for extra virgin olive oil. This points to a niche market for extra virgin olive oil in deep frying.

We report, for the very first time, that the formation of total polar compounds during olive oil deep frying is decreased through sesame oil blending, pointing to a need for further research.

Olive oil is very much appreciated in the Mediterranean and Greek diets, with the biggest benefit coming from phenolic antioxidants. Sesame oil addition did not have any effect on protecting olive oil phenolics during frying. Our study showed that olive oil lost all phenolics after 12 h, while extra virgin olive oil retained around 30%. This result could evolve after further research to a guideline on the use of extra virgin olive oil for deep frying so that the beneficial phenolics are not depleted. Such a guideline could help branding extra virgin olive oil as a health-promoting frying oil. We envisage that deep frying with extra virgin olive oil should not continue until the total polar compounds reach 25%, but should stop much earlier to spare phenolic antioxidants.

Supplementary Materials: The following supporting information can be downloaded at: https://www.mdpi.com/article/10.3390/molecules28114290/s1, Figure S1: TEAC (mmol/Kg) values during deep frying (a) OO, (b) EVOO and the blends with SO; Figure S2: Changes in total phenols (mg GA/Kg oil) during deep frying OO and the blends with SO; Figure S3: Changes in total phenols (mg GA/Kg oil) during deep frying EVOO and the blends with SO; Figure S4: HPLC chromatogram of a mixed standard solution of sesamol, sesamin, and sesamolin; Figure S5: HPLC chromatogram of blend of 20% *v*/*v* sesame oil and olive oil before frying; Figure S6: HPLC chromatogram of blend of 20% *v*/*v* sesame oil and olive oil after 30 min frying; Figure S7: HPLC chromatogram of blend of 20% *v*/*v* sesame oil and olive oil after 1 h of frying; Figure S8: HPLC chromatogram of blend of 20% *v*/*v* sesame oil and olive oil after 2 h of frying; Figure S9: HPLC chromatogram of blend of 20% *v*/*v* sesame oil and olive oil after 4 h of frying; Figure S10: mass spectrum of sesamolin; Figure S11: 1H-NMR spectrum of sesamolin; Figure S12: 1H-NMR spectrum of sesamin; Table S1: total polar compounds during deep frying with olive oil (OO), extra virgin olive oil (EVOO), and blends with sesame oil (SO); Table S2: The retention time, standard curve, LOD, and LOQ for sesame lignans.

Author Contributions: Conceptualization, C.A.G.; methodology, C.A.G.; software, E.T.I.; validation, C.A.G. and E.Z.; formal analysis, E.T.I. and K.S.G.; investigation, C.A.G. and E.T.I.; resources, C.A.G.; data curation, C.A.G. and E.T.I.; writing—original draft preparation, C.A.G. and E.T.I.; writing—review and editing C.A.G. and E.T.I.; visualization, E.Z. and E.T.I.; supervision, C.A.G.; project administration, C.A.G.; funding acquisition, C.A.G. All authors have read and agreed to the published version of the manuscript.

Funding: This research was funded through 'FoodOmics.GR Comprehensive Characterization of Foods' (MIS 5029057) which is implemented under the Action 'Reinforcement of the Research and Innovation Infrastructure', funded by the Operational Programme 'Competitiveness, Entrepreneurship and Innovation' (NSRF 2014–2020), and cofinanced by Greece and the European Union (European Regional Development Fund).

Institutional Review Board Statement: Not applicable.

Informed Consent Statement: Not applicable.

Data Availability Statement: Data will be made available on request.

Acknowledgments: Georgiou acknowledges support for this work from the project 'FoodOmics.GR Comprehensive Characterization of Foods' (MIS 5029057) which is implemented under the Action 'Reinforcement of the Research and Innovation Infrastructure', funded by the Operational Programme 'Competitiveness, Entrepreneurship and Innovation' (NSRF 2014–2020), and cofinanced by Greece and the European Union (European Regional Development Fund).

Conflicts of Interest: The authors declare no conflict of interest.

Sample Availability: Not applicable.

References

1. Zhang, Q.; Saleh, A.; Chen, J.; Shen, Q. Chemical alterations taken place during deep-fat frying based on certain reaction products: A review. *Chem. Phys. Lipids* **2012**, *165*, 662–681. [CrossRef] [PubMed]
2. Choe, E.; Min, D.B. Chemistry of deep-fat frying oils. *J. Food Sci.* **2007**, *72*, R77–R78. [CrossRef] [PubMed]
3. Nayak, P.K.; Dash, U.; Rayaguru, K.; Krishnan, K.R. Physio-chemical changes during repeated frying of cooked oil: A review. *J. Food Biochem.* **2016**, *40*, 371–390. [CrossRef]
4. Dobarganes, C.; Marquez-Ruiz, G. Possible adverse effects of frying with vegetable oils. *Br. J. Nutr.* **2015**, *113*, S49–S57. [CrossRef] [PubMed]
5. Hosseini, H.; Ghorbani, M.; Meshginfar, N.; Mahoonak, A.S. A Review on frying: Procedure, fat, deterioration progress and health hazards. *J. Am. Oil Chem. Soc.* **2016**, *93*, 445–466. [CrossRef]
6. Romani, A.; Leri, F.; Urciuoli, S.; Noce, A.; Marrone, G.; Nediani, C.; Bernini, R. Health effects of phenolic compounds found in extra-virgin olive oil, by-products, and leaf of *Olea europaea* L. *Nutrients* **2019**, *11*, 1776. [CrossRef]
7. Lozano-Castellon, J.; de Alvarenga, J.F.R.; Vallverdu-Queralt, A.; Lamuela-Raventos, R.M. Cooking with extra-virgin olive oil: A mixture of food components to prevent oxidation and degradation. *Trends Food Sci. Technol.* **2022**, *123*, 28–36. [CrossRef]
8. Visioli, F.; Franco, M.; Toledo, E.; Luchsinger, J.; Willett, W.C.; Hu, F.B.; Martinez-Gonzalez, M.A. Olive oil and prevention of chronic diseases: Summary of an International conference. *Nutr. Metab. Cardiovas. Dis.* **2018**, *28*, 649–656. [CrossRef]
9. Wani, T.A.; Masoodi, F.A.; Gani, A.; Baba, W.N.; Rahmanian, N.; Akhter, R.; Wani, I.A.; Ahmad, M. Olive oil and its principal bioactive compound: Hydroxytyrosol—A review of the recent literature. *Trends Food Sci. Technol.* **2018**, *77*, 77–90. [CrossRef]
10. Casal, S.; Malheiro, R.; Sendasa, A.; Oliveira, B.P.; Pereira, J.A. Olive oil stability under deep frying conditions. *Food Chem. Toxicol.* **2010**, *48*, 2972–2979. [CrossRef]
11. Wua, G.; Changa, C.; Honga, C.; Zhanga, H.; Huanga, J.; Jina, Q.; Wanga, X. Phenolic compounds as stabilizers of oils and antioxi-dative mechanisms under frying conditions: A comprehensive review. *Trends Food Sci. Technol.* **2019**, *92*, 33–45. [CrossRef]
12. Farhoosh, R.; Kenari, R.E. Anti-rancidity effects of sesame and rice bran oils on canola oil during deep frying. *J. Am. Oil Chem. Soc.* **2009**, *86*, 539–544. [CrossRef]
13. Farhoosh, R.; Tavassoli-Kafrani, M.H.; Sharif, A. Assaying antioxidant characteristics of sesame seed, rice bran, and bene hull oils and their unsaponifiable matters by using DPPH radical-scavenging model system. *J. Agric. Sci. Technol.* **2013**, *15*, 241–253.
14. Namiki, M. The chemistry and physiological functions of sesame. *Food Rev. Int.* **1995**, *11*, 281–329. [CrossRef]
15. Andargie, M.; Vinas, M.; Rathgeb, A.; Möller, E.; Karlovsky, P. Lignans of Sesame (*Sesamum indicum* L.): A Comprehensive Review. *Molecules* **2021**, *26*, 883. [CrossRef]
16. Fukuda, Y.; Nagata, M.; Osawa, T.; Namiki, M. Chemical Aspects of the antioxidative activity of roasted sesame seed oil, and the effect of using the oil for frying. *Agric. Biol. Chem.* **1986**, *50*, 857–862.
17. Suja, K.P.; Jayalekshmy, A.; Arumughan, C. Free radical scavenging behavior of antioxidant compounds of sesame (*Sesamu-mindicum* L.) in DPPH system. *J. Agric. Food Chem.* **2004**, *52*, 912–915. [CrossRef]

18. Lee, S.W.; Jeung, M.K.; Park, M.H.; Lee, S.Y.; Lee, J.H. Effects of roasting conditions of sesame seeds on the oxidative stability. *Food Chem.* **2010**, *118*, 681–685. [CrossRef]
19. Hemalatha, S.; Ghafoorunissa. Sesame lignans enhance the thermal stability of edible vegetable oils. *Food Chem.* **2007**, *105*, 1076–1085. [CrossRef]
20. Lee, J.; Lee, Y.; Choe, E. Effects of sesamol, sesamin, and sesamolin extracted from roasted sesame oil on the thermal oxidation of methyl linoleate. *LWT—Food Sci. Technol.* **2008**, *41*, 1871–1875. [CrossRef]
21. Borjian Borojeni, M.; Goli, S.A.H.; Gharachourloo, M. Effect of roasted sesame oil on qualitative properties of frying oil during deep-fat frying. *J. Agric. Sci. Technol.* **2016**, *18*, 1531–1542.
22. Weisshaar, R. Quality control of used deep frying oils. *Eur. J. Lipid Sci. Technol.* **2014**, *116*, 716–722. [CrossRef]
23. Doborganes, M.C.; Velasco, J.; Dieffenbacher, A. Determination of polar compounds polymerized and oxidized triacyl-glycerol and diacylglycerols in oils and fats. *Pure Appl. Chem.* **2000**, *72*, 1563–1575. [CrossRef]
24. Paul, S.; Mittal, G.S. Regulating the use of degraded oil/fat in deep-fat/oil food frying. *Crit. Rev. Food Sci.* **1997**, *37*, 635–662. [CrossRef] [PubMed]
25. Fukuda, Y.; Osawa, T.; Kawakishi, S.; Namiki, M. Chemistry of lignan antioxidants in sesame seed and oil. In *Food Phytochemicals for Cancer Prevention II, Teas, Spices and Herbs*; ACS Symposium Series 547; American Chemical Society: Washington, DC, USA, 1994; Chapter 27, pp. 264–274.
26. Chantzos, N.V.; Georgiou, C.A. Monitoring lipid oxidation events at frying temperatures through radical scavenging assays. *Chem. Ind. Chem. Eng. Q.* **2007**, *13*, 163–166. [CrossRef]
27. Hemalatha, S.; Ghafoorunissa. Lignans and tocopherols in Indian sesame cultivars. *J. Am. Oil Chem. Soc.* **2004**, *81*, 467–470. [CrossRef]
28. Yeo, J.D.; Park, J.W.; Lee, J.H. Evaluation of antioxidant capacity of sesamol and free radical scavengers at different heating temperatures. *Eur. J. Lipid Sci. Technol.* **2011**, *113*, 910–915. [CrossRef]
29. Hwang, H.; Winkler-Moser, J. *Oxidative Stability and Shelf Life of Foods Containing Oils and Fats*; Elsevier: London, UK; AOCS American Oil Chemists' Society: Urbana, IL, USA, 2016; pp. 251–285.
30. Bazina, N.; He, J. Analysis of fatty acid profiles of free fatty acids generated in deep-frying process. *J. Food Sci. Technol.* **2018**, *55*, 3085–3092. [CrossRef]
31. Rodríguez, G.; Giacomo Squeo, G.; Estivi, L.; Quezada Berru, S.; Buleje, D.; Caponio, F.; Brandolinie, A.; Hidalgo, A. Changes in stability, tocopherols, fatty acids and antioxidant capacity of sacha inchi (*Plukenetia volubilis*) oil during French fries deep-frying. *Food Chem.* **2011**, *340*, 127942. [CrossRef]
32. Kalantzakis, G.; Blekas, G.; Pegklidou, K.; Boskou, D. Stability and radical-scavenging activity of heated olive oil and other vegetable oils. *Eur. J. Lipid Sci. Technol.* **2006**, *108*, 329–335. [CrossRef]
33. Jeong, M.K.; Yeo, J.D.; Jang, E.Y.; Kim, M.J.; Lee, J.H. Aldehydes from oxidized lipids can react with 2,2-diphenyl-1-picrylhydrazyl (DPPH) free radicals in isooctane systems. *J. Am. Oil Chem. Soc.* **2012**, *89*, 1831–1838. [CrossRef]
34. Lopez, C.T.; Carpena, M.; Lourenço-Lopes, C.; Gallardo-Gomez, M.; Lorenzo, J.M.; Barba, F.J.; Prieto, M.A.; Simal-Gandara, J. Bioactive compounds and quality of extra virgin olive oil. *Foods* **2020**, *9*, 1014. [CrossRef]
35. Boskou, D.; Blekas, G.; Tsimidou, M. Olive Oil Composition. In *Olive Oil: Chemistry and Technology*; AOCS Press: New York, NY, USA, 2006; Chapter 4, pp. 41–72.
36. *Codex Stan 210-1999*; Standard for Named Vegetable Oils. Joint FAO/WHO Codex Alimentarius Commission: Rome, Italy, 2013.
37. Wu, W.-H. The contents of lignans in commercial sesame oils of Taiwan and their changes during heating. *Food Chem.* **2007**, *104*, 341–344. [CrossRef]
38. European Union. *REGULATION (EEC) No 2568/91 of 11 July 1991 on the Characteristics of Olive Oil and Olive-Residue Oil and on the Relevant Methods of Analysis*; European Union: Maastricht, The Netherlands, 1991.
39. Capannesi, C.; Palchetti, I.; Mascini, M.; Parenti, A. Electrochemical sensor and biosensor for polyphenols detection in olive oils. *Food Chem.* **2000**, *71*, 553–562. [CrossRef]

molecules

MDPI

Article

Detection of Soybean-Derived Components in Dairy Products Using Proofreading Enzyme-Mediated Probe Cleavage Coupled with Ladder-Shape Melting Temperature Isothermal Amplification (Proofman–LMTIA)

Fugang Xiao [1,*,†], Menglin Gu [1,2,†], Yaoxuan Zhang [1,2,†], Yaodong Xian [1], Yaotian Zheng [1], Yongqing Zhang [1], Juntao Sun [1], Changhe Ding [2], Guozhi Zhang [2] and Deguo Wang [1,*]

1 Henan Key Laboratory of Biomarker Based Rapid-Detection Technology for Food Safety, Food and Pharmacy College, Xuchang University, Xuchang 461000, China
2 College of Grain and Food, Henan University of Technology, Zhengzhou 450001, China
* Correspondence: xfug@163.com (F.X.); wangdg666@126.com (D.W.); Tel.: +86-374-2968805 (F.X.); +86-374-2968907 (D.W.)
† These authors contributed equally to this work.

Abstract: Food adulteration is a serious problem all over the world. Establishing an accurate, sensitive and fast detection method is an important part of identifying food adulteration. Herein, a sequence-specific ladder-shape melting temperature isothermal amplification (LMTIA) assay was reported to detect soybean-derived components using proofreading enzyme-mediated probe cleavage (named Proofman), which could realize real-time and visual detection without uncapping. The results showed that, under the optimal temperature of 57 °C, the established Proofman–LMTIA method for the detection of soybean-derived components in dairy products was sensitive to 1 pg/μL, with strong specificity, and could distinguish soybean genes from those of beef, mutton, sunflower, corn, walnut, etc. The established Proofman–LMTIA detection method was applied to the detection of actual samples of cow milk and goat milk. The results showed that the method was accurate, stable and reliable, and the detection results were not affected by a complex matrix without false positives or false negatives. It was proved that the method could be used for the detection and identification of soybean-derived components in actual dairy products samples.

Keywords: food adulteration; dairy products; ladder-shape melting temperature isothermal amplification (LMTIA); soybean-derived components; proofreading enzyme-mediated probe cleavage (Proofman)

Citation: Xiao, F.; Gu, M.; Zhang, Y.; Xian, Y.; Zheng, Y.; Zhang, Y.; Sun, J.; Ding, C.; Zhang, G.; Wang, D. Detection of Soybean-Derived Components in Dairy Products Using Proofreading Enzyme-Mediated Probe Cleavage Coupled with Ladder-Shape Melting Temperature Isothermal Amplification (Proofman–LMTIA). *Molecules* **2023**, *28*, 1685. https://doi.org/10.3390/molecules28041685

Academic Editor: Gavino Sanna

Received: 27 December 2022
Revised: 28 January 2023
Accepted: 6 February 2023
Published: 10 February 2023

1. Introduction

In recent years, with the rapid development in China's economy and the improvement in people's living standards, the demands for food quality and safety have also improved. At the same time, dairy products as a kind of food with high nutritional value have become an indispensable food in the diet structure of residents in our country. With the increase in people's income and change in consumption concept, consumers in the dairy market continue to pursue healthy, high-quality and safe products, and promote the continuous upgrading of dairy consumption structure [1]. According to the data from 2005 to 2017, the data envelopment analysis (DEA) model was used to analyze the quality of dairy products in different importing countries due to different production technologies [2], and there was the problem that importing countries were too concentrated [3]. Therefore, it is necessary to establish an efficient and rapid detection technology for dairy products.

Driven by profits, fraud such as shoddy and adulterated dairy products with high value often occurs [4–6]. At present, the main adulterants in dairy products are nitrogen-containing compounds, vegetable proteins (such as soy protein) and animal proteins. Water, rice soup, hydrolyzed leather protein, plant fat end, soy protein, whey powder, melamine and urea are usually added to dairy products [7]. After the "Big Head Baby" scandal of Sanlu cow milk powder [8], with the efforts of many scientific researchers, the detection methods of melamine and other nonprotein nitrogen content have become more and more advanced, while profitmaking through the adulteration of nonprotein nitrogen content has become a thing of the past. However, low-cost plant-derived proteins have been used to replace the nonprotein nitrogen, and the main plant-derived proteins are soybean proteins [9]. Therefore, in the process of quality control, which is particularly important for government supervision and enterprise production, it is necessary to use fast and reliable detection methods to monitor raw materials.

Due to the continuous improvement in adulteration technology, it is difficult to identify the authenticities in dairy products using ordinary sensory methods or simple chemical methods [10–13]. DNA-based detection methods are highly sensitive and repeatable, and have been widely used in the adulteration of dairy products [14].

The nucleic acid detection method has strong specificity, high sensitivity and accurate results [15] and is suitable for species identification [16]. The existing nucleic acid amplification methods can be divided into two categories: thermal cycle amplification and isothermal amplification [17]. The commonly used thermal cycle amplification methods are polymerase chain reaction (PCR) and ligase chain reaction (LCR) [18–20]. The limitation of these methods is that they need a thermal cycle instrument. Isothermal amplification technology is developed on the basis of thermal cycle amplification technology. Isothermal amplification does not need a thermal circulator and is simpler and shorter than variable-temperature amplification. There are many technologies that have been developed, such as nucleic acid sequence-based amplification (NASBA) [21], self-sustained sequence replication (3SR) [22], helicase-dependent isothermal DNA amplification (HDA) [23], exponential amplification reaction (EXPAR) [24], strand displacement amplification (SDA) [25], recombinase polymerase amplification (RPA) [26], cross-primer amplification (CPA), rolling circle amplification (RCA) [27], loop-mediated isothermal amplification (LAMP) [28]. However, these methods still have different disadvantages such as a long amplification time and nonspecific amplification [28,29].

The DNA mentioned in the LAMP principle is in dynamic equilibrium at about 65 °C. When any primer extends to the complementary part of the double-stranded DNA for base pairing, the other strand will dissociate and become a single strand. If this theory is established, one pair of primers can also be amplified, so there is no need for high-temperature denaturation in the PCR reaction. Adding helicase to the constant-temperature amplification technology of DNA, dependent on helicase to unlock the double strand of DNA, will also be superfluous. Our team has proved that this theory is wrong [30,31]. However, in many literature reports, based on this theoretical basis, the reaction was realized through LAMP. Our team analyzed this and believed that the single strand of DNA was generated due to the influence of heating or chemical factors during the preparation of the DNA template, and the amplification was realized, which further increased the chance and uncertainty of this amplification. If the 3–4 bases at the 3′ end of any two primers of LAMP have two complementary sequences on the same primer, there must be nonspecific amplification under the common LAMP reaction system and reaction conditions. The design and screening of LAMP primers should avoid this situation, but the mechanism of this nonspecific amplification needs further study. Taking the specific gene *inv*A of *Salmonella* as the target sequence, a set of LAMP primers was designed and screened, which did not exist in this case, and the negative control did not have nonspecific amplification. In short, the false positive in the LAMP reaction is mainly caused by nonspecific amplification between primers; thus, primer design and screening are crucial [30,31]. Based on the above analysis, with reference to denaturing gradient gel electrophoresis (DGGE), our team improved the method of micro-denaturing partial chain breaking and independently developed a new isothermal amplification technology with only one pair of primers: ladder-shape melting temperature isothermal amplification (LMTIA) [32,33]. Its amplification can be divided into two major categories, namely, herringbone type and nonherringbone type, and subdivided into eight subcategories according to the melting temperature curve and primer type. The LMTIA reaction is divided into the initial structure formation stage and the exponential amplification stage. The LMTIA technology uses one pair of primers or two pairs of nested primers and a thermostable DNA polymerase (large fragment) to amplify the internal transcribed spacer (ITS) of rice in a ladder melting curve. Compared with the LAMP method with the same level of specificity, the sensitivity of the nested-primer LMTIA method is increased by 50 times. The primer design method of LMTIA is simple, and the amplification efficiency of LMTIA is high and fast. The amplification reaction enters the platform stage from the exponential stage in about 20 min.

The LMTIA technology makes the thermal cycler no longer a necessary instrument for nucleic acid amplification technology [34]. The LMTIA technology has the following advantages [32]:

(1) the reaction mechanism is simple and it does not need thermal denaturation, nor does it need the synergy of other enzymes except Bst DNA polymerase;

(2) the primer design is simple, and the LMTIA primers can be designed using common PCR primer software;

(3) the reaction time is short, and the amplification reaction time is less than 30 min;

(4) the specificity is strong, because the LMTIA reaction requires not only the complementary pairing of the primer and the target sequence, but also the ladder structure of the melting temperature curve of the target sequence;

(5) it has high sensitivity, reported to be more than 50 times higher than the LAMP;

(6) low requirements for the length of the target sequence. LMTIA can amplify more than 60 nt of the target sequence, while LAMP requires the length of the target sequence to be at least 200 nt;

(7) it can be used in a wide range of applications, including in double-stranded DNA as a template or in single-stranded DNA or RNA as a template, and does not require reverse transcriptase [35].

Compared with the traditional PCR technology, it is accurate for the objective and it has a relatively low detection limit and short amplification time. After the amplification, the fluorescence generation can be directly observed. Currently, it is widely used in the detection of food adulteration [36,37]. A rapid method for the detection of dairy soybean genes was established, which can be used for the primary screening of dairy products.

Probes based on enzymatic cleavage have a wide range of applications in molecular biology, but their preparation is complex and requires the incorporation of tetrahydrofuran base-free site analogues. Chen et al. cleaved the enzyme mediated probe (Proofman) and developed recombinase-assisted loop-mediated isothermal amplification (RALA) [38,39]. A novel nucleic acid sequence-specific detection platform for soybean was developed using LMTIA coupled with a Proofman probe. The Proofman-based LMTIA was proved to be an attractive option for accurately diagnosing soybean. In this study, the nucleic acid sequence of soybean (*Glycine max* (Linn.) Merr.) endogenous gene was selected to design the LMTIA specific primers, and a specific, rapid and accurate LMTIA-coupled Proofman detection method was established to detect soybean-derived components in dairy products.

2. Results

2.1. LMTIA Primer Design

The primer sequences are shown in Table 1. Then, using the Proofman probe, labeled with a fluorophore and quencher at the 3′ end and 5′ end, respectively, the sequence-specific detection of LMTIA products was achieved through the binding and cleavage of the Proofman probe, thus enhancing the accuracy of the LMTIA reaction.

Table 1. LMTIA primer sequences for ITS gene in soybean.

Primer	Sequence (5′–3′)
F	CGTGCACGCAAAGGGTTTTTCCACGCTCGAGACCAATCAC
B	TGCACGCACGCTCCCTTTTATGCTTAAACTCAGCGGGTAG
LF	TCCAGAACTGACCGGCTCGCA
LB	ACGAGACCTCAGGTCAGGCG
Pr	BHQ 1-ACGAGACCTCAGGTCAGGCG-FAM

2.2. Optimization of the Proofman–LMTIA Reaction Temperature

The Proofman–LMTIA assay was conducted at 55 °C, 56 °C, 57 °C and 58 °C, and the results are shown in Figure 1. The negative control DEPC water was not amplified, while the soybean gDNA was amplified and the amplification effect was good. Soybean primers were not significantly affected by temperature change and began to amplify at around 12 cycles, but from the perspective of fluorescence intensity and repeatability, 57 °C was the best, so it was determined as the best reaction temperature.

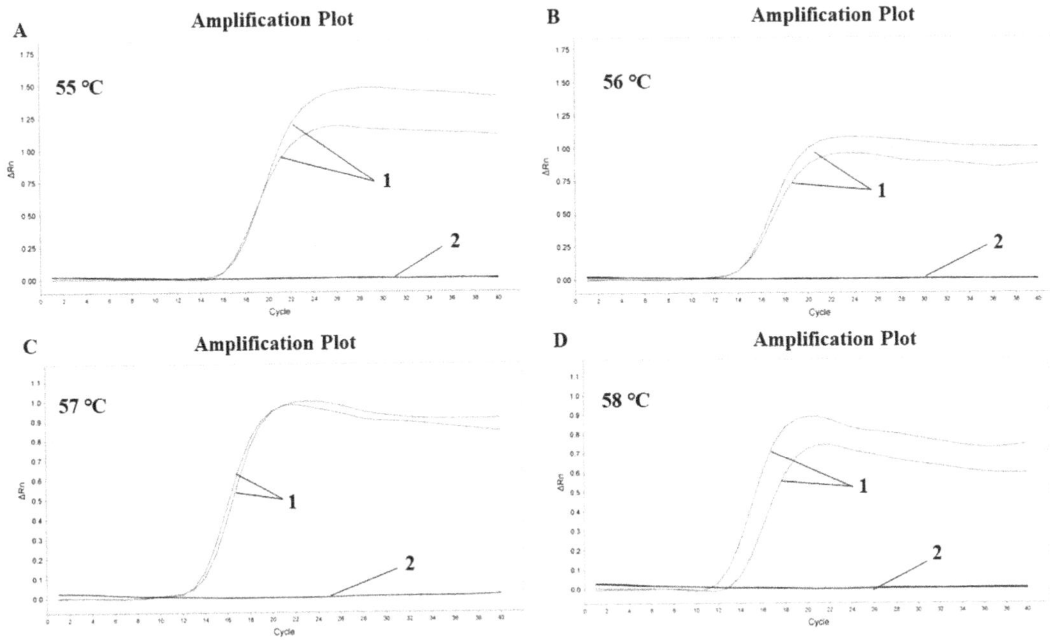

Figure 1. Optimization of Proofman–LMTIA reaction temperature for ITS gene in soybean. (**A**) 55 °C, (**B**) 56 °C, (**C**) 57 °C, (**D**) 58 °C. 1: positive controls (10 ng of gDNA from soybean); 2: negative controls (ddH_2O).

2.3. Specificity of the Proofman–LMTIA Assay

In order to ensure the unique specificity of the designed primers for soybean, the optimized Proofman–LMTIA reaction system was used to perform a Proofman–LMTIA test on the gDNA grouping of soybean, sesame, sunflower, walnut, sheep, cattle and others, with DEPC water as the negative control and soybean gDNA as the positive control. The result is shown in Figure 2. Under the condition of 57 °C, soybean DNA was amplified normally, while DEPC water and other control DNA were not amplified. Therefore, the Proofman–LMTIA assay was of high specificity and only amplified the soybean DNA, and the primers designed in this experiment had good specificity and could meet the screening of soybean-derived components in dairy products.

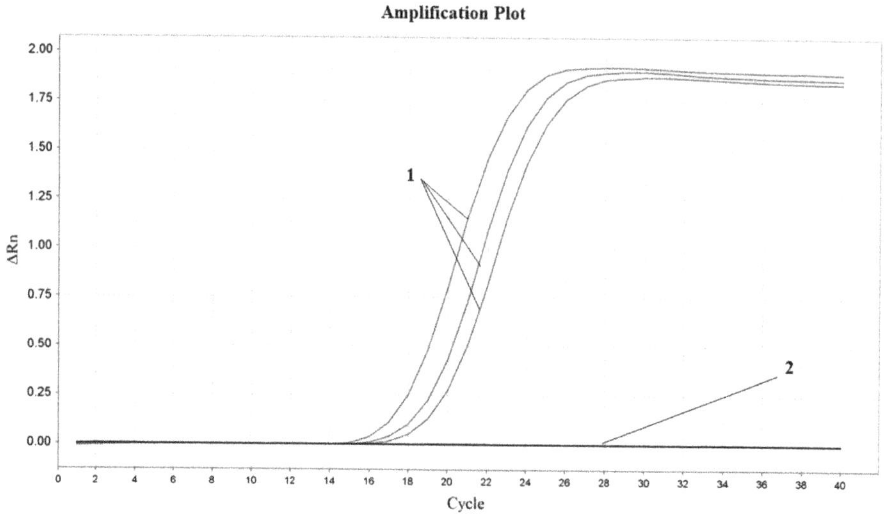

Figure 2. Proofman–LMTIA response specificity assay of ITS gene in soybean. 1: Soybean gDNA; 2: ddH$_2$O, sesame, sunflower, walnut, sheep, cattle and others.

2.4. Sensitivity of the Proofman–LMTIA Assay

In order to ensure the sensitivity of the experimental method, the sensitivity of the designed soybean primer was tested. The cryopreserved soybean DNA was diluted to three concentration gradients of 100 pg/μL, 10 pg/μL and 1 pg/μL, respectively, and the Proofman–LMTIA assay was performed at 57 °C with DEPC water as the negative control and soybean DNA as the positive control. The amplification curve is shown in Figure 3. Soybean DNA at concentrations of 100 pg/μL, 10 pg/μL and 1 pg/μL were all well amplified, but the stability and repeatability were not good with 1 pg/μL. Therefore, the concentration of soybean DNA detected using the Proofman–LMTIA assay was 1 pg/μL.

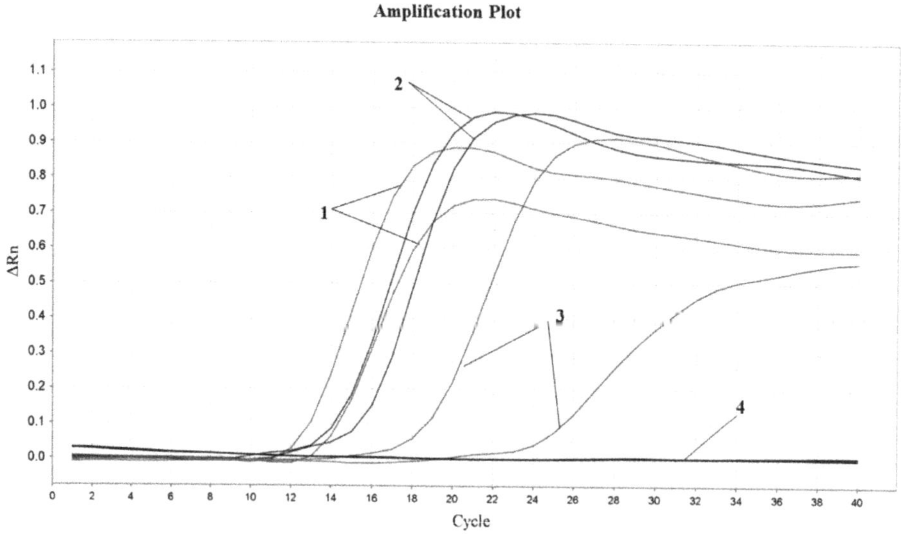

Figure 3. Proofman–LMTIA response sensitivity test for ITS gene in soybean. Varying amounts of gDNAs of the following: 1: soybean 100 pg/μL; 2: soybean 10 pg/μL; 3: soybean 1 pg/μL; 4: ddH$_2$O.

2.5. Sample Testing

In order to verify whether the system could detect soybean-derived components in dairy products, Mengniu pure cow milk and Ama-jia pure cow milk of the Qinghai-Tibet Plateau were tested, respectively, under the soybean reaction system and the bovine reaction system. In the bovine reaction system, DEPC water was the negative control and bovine DNA was the positive control. Both cow milk and cow DNA were amplified. In the soybean reaction system, DEPC water was also used as the negative control and soybean DNA was used as the positive control. The cow milk DNA was unamplified, while the soy DNA was amplified. The results are shown in Figures 4 and 5. The same method was used to detect Jomilk selected pure goat milk, and the results are shown in Figure 6. In the sheep system, the DNA from both sheep milk and goat milk was amplified. In the soy system, only soy DNA was amplified.

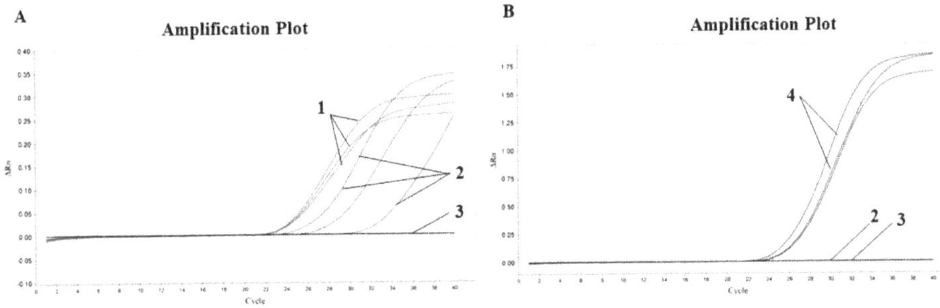

Figure 4. Proofman–LMTIA method for testing Mengniu pure cow milk. (**A**): Verification of cow milk DNA: 1: beef gDNA; 2: cow milk DNA; 3: ddH$_2$0; (**B**): Detection of soybean-derived components in cow milk: 2: cow milk DNA; 3: ddH$_2$0; 4: soybean gDNA.

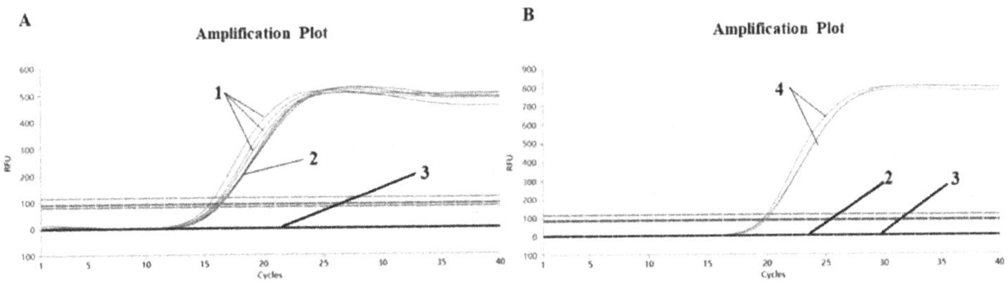

Figure 5. Proofman–LMTIA method for testing Ama's pure cow milk in Qinghai–Tibet Plateau. (**A**): Verification of cow milk DNA: 1: beef gDNA; 2: cow milk DNA; 3: ddH$_2$0; (**B**): Detection of soybean-derived components in cow milk: 2: cow milk DNA; 3: ddH$_2$0; 4: soybean gDNA.

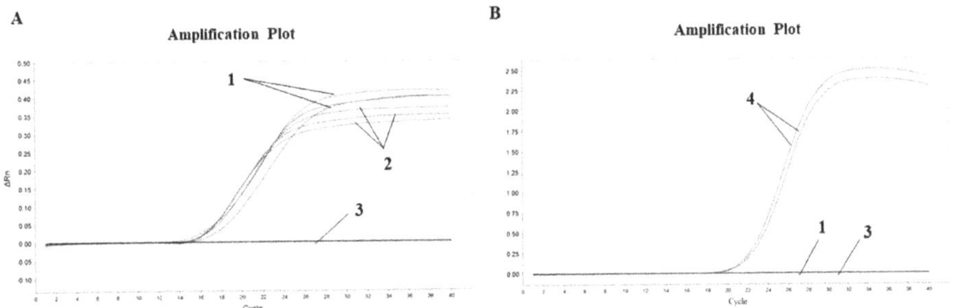

Figure 6. Proofman–LMTIA method for detection of Jomilk selected pure goat milk. (**A**): Verification of goat milk DNA: 1: goat milk DNA; 2: mutton gDNA; 3: ddH$_2$0; (**B**): Detection of soybean-derived components in goat milk: 1: goat milk DNA; 3: ddH$_2$0; 4: soybean gDNA.

3. Material and Methods

3.1. Target Sequence Selection and LMTIA Primer Design

There are three criteria for the target sequence selection, as described by Wang et al. [32]: the melting temperature curve of the sequence is of a ladder type; the GC content of the sequence is generally 40–80%; and the sequence has high specificity. The sequence with a ladder-shaped melting temperature curve was selected as a target from the internal transcribed spacer (ITS) gene of soybean using the software Oligo 7 (Molecular Biology Insights, Inc. Colorado Springs, CO, USA) [32]. The LMTIA primers were designed with the online software Primer3Plus (http://www.primer3plus.com, accessed on 20 March 2022), and the parameters of primers were set as described in a previous report [32]. The Proofman probes were designed based on the primer LB sequence, with the fluorophore and the quencher labeled at the end of the 3′ end mismatch nucleotide and 5′ end nucleotide, respectively [39].

3.2. DNA Extraction

When extracting DNA from cow milk and goat milk, DNA enrichment is required first, and the specific steps are as follows: 10 mL milk samples were collected in a centrifugal tube and centrifuged at 4 °C at 7000 r/min for 10 min. Then, 600 μL of PBS was added to the bottom precipitate, and the precipitate was suspended by repeated blowing and transferred to a 2 mL centrifuge tube. Then, the precipitate was put into a high-speed refrigerated centrifuge and centrifuged at 4 °C and 12,000 r/min for 10 min; the supernatant was poured out; and the bottom precipitate was removed. After the bottom precipitation was weighed, DNA was extracted from the corresponding milk samples using a food DNA extraction kit (TIANGEN DP326, Tiangen Biotech [Beijing] Co., Ltd., Beijing, China). After the purity and concentration were determined, the samples were frozen at 4 °C.

A plant genome kit (TIANGEN DP305, Tiangen Biotech [Beijing] Co., Ltd., Beijing, China) was used to extract DNA from soybean, sunflower, sesame, walnut and other plants, and an animal genome kit (TIANGEN DP304, Tiangen Biotech [Beijing] Co., Ltd., Beijing, China) was used to extract DNA from beef and mutton. After the DNA purity and concentration were determined, the DNA was stored at 4 °C.

3.3. Proofman–LMTIA Reaction

For the Proofman–LMTIA reaction, 10 μL of the reaction system (0.06 μL of primer F and primer B, 0.02 μL of primer LF and primer LB, 0.014 μL of Pr, 4 μL of the premix of dNTP and Bst DNA Polymerase (Merit Biotech [Shandong] Co., Ltd., Heze, China), and 2 μL of DNA template) was added for the conventional fluorescent reaction. As for Proofman, instead of SYBR Green I, 0.014 μL of the Proofman probe was added.

3.4. Proofman–LMTIA Reaction Temperature Optimization

According to the reaction system described in 3.3, the Stepone plus real-time PCR system (Applied Biosystems, Thermo Fisher Scientific, Foster City, CA, USA) reaction temperatures were set as 55 °C, 56 °C, 57 °C and 58 °C, respectively, with heating for 60 min (90 s for 1 cycle; the purpose of the setting cycle was 90 s for the one-time acquisition of the fluorescence signal and FAM was selected for the fluorescence channel) [10]. The positive control was soybean genomic DNA (gDNA), and the negative control was DEPC water. After the amplification, the amplification curve of real-time fluorescent quantitative Proofman–LMTIA was carefully observed and studied to determine the optimal reaction temperature of soybean primers.

3.5. Specificity Determination of Proofman–LMTIA Assay

With H_2O as the negative control and 1 ng/μL soybean gDNA as a positive control, the specificity of the established Proofman–LMTIA assay was tested with the gDNAs of beef, mutton, sunflower, corn, walnuts and others. A 10 μL reaction system described in 3.3 was applied, and each sample was repeated twice (repeated thrice for soybean gDNA) in the Archimed time-resolved fluorescence quantitative PCR system (RocGene [Beijing] Technology Co., Ltd., Beijing, China).

3.6. Sensitivity Determination of Proofman–LMTIA Assay

Soybean gDNA was diluted to 100 pg/μL, 10 pg/μL and 1 pg/μL, and the sensitivity was tested under the optimal temperature reaction conditions. A 10 μL reaction system described in 3.3 was applied, and two parallel samples were set for each sample.

3.7. Actual Sample Testing

The gDNA of a brand of cow milk and goat milk was extracted and verified using the previously established Proofman–LMTIA method of beef and mutton [40,41]. Then, the established soybean Proofman–LMTIA detection system was used to detect whether soybean-derived components were present. DEPC water was the negative control and beef gDNA was the positive control; in the sheep reaction system, DEPC water was the negative control and mutton gDNA was the positive control; and in the soybean reaction system, DEPC water was the negative control and soybean gDNA was the positive control. A reaction system in an amount of 10 μL was used with at least two parallelisms for each sample.

The flow chart of Proofman–LMTIA is shown in Figure 7.

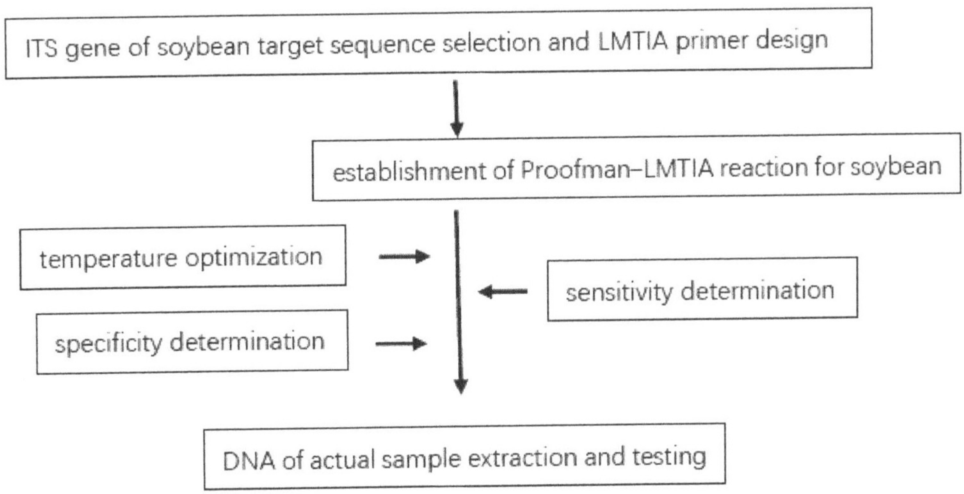

Figure 7. Flow chart of Proofman–LMTIA.

4. Discussion

The separation and extraction of high-quality DNA is the key to genetic testing. DNA extracted from dairy products is mainly obtained from somatic cells, but there are only 20,000 to 200,000 individual cells per milliliter of milk, and the DNA content is not high [42]. In addition, dairy products need to go through spray drying, high-temperature sterilization and other processing in the production process, which will lead to DNA breakage or degradation, exacerbating the difficulty of DNA extraction. In this study, DNA extraction kits suitable for dairy products and other different samples were selected through comparative tests; the amplified DNA from dairy products was successfully extracted using nucleic acid enrichment and a food DNA extraction kit.

As reported in the previous paper [32], the LMTIA technique based on LAMP can be used under isothermal conditions to detect *Oryza sativa* L. DNA with 50 times more sensitivity than LAMP. The sequence with a ladder-shaped melting temperature curve was selected as a target from the internal transcribed spacer (ITS) gene of soybean using the software Oligo 7. Based on the selected sequence, LMTIA primers were designed, and meanwhile, the Proofman probes were also designed based on the primer LB sequence, with the fluorophore and the quencher labeled at the end of the 3′ end mismatch nucleotide and the 5′ end nucleotide, respectively. Further, the primers displayed high specificity at 57 °C for the soybean DNA (Figure 1). This simple and rapid method could be applied to determine the soybean-derived components in dairy products.

In recent years, food fraud has become an urgent problem linked to the traceability of food products, and food authenticity has become increasingly important as a result of food adulteration [37]. Dairy products are one of the important sources of nutrition for the human body. However, fraud—driven by profits—often occurs, such as shoddy and adulterated dairy products with high value [4–6]. Therefore, the development of selective and sensitive detection techniques is a key challenge for the detection of authentic food, especially dairy products. DNA-based detection methods are highly sensitive and repeatable, and have been widely used in the adulteration of food products [14]. Zhu et al. (2019) [36] reported that the LAMP method could be used to detect pork components in common meat products, and the detection limit was 10 pg/uL. The LMTIA assay was also a reliable method for the rapid detection of cassava components in sweet potato starch noodles and could specifically distinguish a 0.01% (*w*/*w*) cassava component added into sweet potato starch [37]. Wang et al. (2022) [40] developed the LMTIA method for the detection of duck adulteration in beef with a 0.1% limit of detection. The established Proofman–LMTIA method for the detection of soybean-derived components in dairy products was sensitive to 1 pg/μL (Figure 3). Therefore, the DNA-based detection methods can be applied in the detection of authentic food, as well as the preliminary screening of food products on the market. In addition, with the advantages of simple equipment requirements, simple operation, high detection efficiency, rapid detection, strong specificity and high sensitivity, it is suggested to continuously strengthen the integrity and application of Proofman–LMTIA technology, and explore its application in food adulteration detection, pathogenic bacteria detection, virus detection and species relationship detection.

In this study, with a Proofman probe instead of SYBR Green I, multiplex detection can be achieved in one pot, improving the accuracy of detection. Multiple probes can be added into the system to realize multiple detection, which is suitable for the rapid detection of food adulteration.

The LMTIA method requires not only the complementary pairing of the primer and target sequence, but also the ladder-type melting curve of the target sequence to provide a single-chain template for the amplification reaction [32]. Therefore, in theory, the LMTIA method should have high specificity. In this study, the established Proofman–LMTIA method was used to detect the gDNA grouping of sesame, sunflower, walnut, sheep, cattle and others, with DEPC water as the negative control, which was negative, and the detected gDNA of soybean as the positive control, which preliminarily confirmed the specificity of the Proofman–LMTIA method.

On the basis of previous research, further research was carried out to clarify the mechanism of template unwinding in LAMP technology through theoretical analysis and practical verification, and to use a pair of primers and other technical means to solve the false-positive problem caused by nonspecific amplification and apply it to the LMTIA technology. The primer design screening method was established using mathematical induction, and the LMTIA technology was improved by combining a fluorescent probe and a catalytic enzyme to carry out the qualitative and quantitative detection of the authenticity of dairy products so that its sensitivity reached the same level as that of fluorescent PCR; the detection time was shortened to 20 min so as to further improve the LMTIA technology. This study can further improve the theory of nucleic acid detection, break through the bottleneck of LMTIA technology and promote the development and application of new rapid diagnoses of epidemic diseases and rapid detection kits for food safety.

5. Conclusions

LMTIA primers were designed for the soybean ITS gene, and a method of Proofman–LMTIA detection of soybean-derived components in dairy products was developed. The specific test showed that the primers could distinguish soybean DNA from the DNA of sesame, sunflower, walnut, mutton and beef. The optimal reaction temperature of the established soybean Proofman–LMTIA assay was 57 °C, and the sensitivity was up to 1 pg/μL at this temperature. The Proofman–LMTIA method established in this study has the advantages of simple equipment requirements, simple operation, high detection efficiency, rapid detection, high sensitivity and strong specificity, and can be applied to the detection of soybean-derived components in dairy products.

Author Contributions: F.X.: Methodology, writing—review and editing and funding acquisition; M.G.: Methodology, data analysis and writing—original draft; Y.Z. (Yaoxuan Zhang): Methodology, material and formal analysis; Y.X. and Y.Z. (Yaotian Zheng): Methodology; Y.Z. (Yongqing Zhang): Visualization; J.S.: Software and formal analysis; C.D. and G.Z.: Supervision and resources; D.W.: Conceptualization, supervision and funding acquisition. All authors have read and agreed to the published version of the manuscript.

Funding: This research was funded by the Henan Provincial Key Research Program for Higher Education Institutions (Grant No. 21B550006) and the Technology Project of Henan Province (Grant No. 212102110214).

Institutional Review Board Statement: Not applicable.

Informed Consent Statement: Not applicable.

Data Availability Statement: The data presented in this study are available on request from the corresponding author.

Acknowledgments: We thank Penghua Shu and Bochun Hu from Xuchang University for their help in reviewing this paper.

Conflicts of Interest: The authors declare no conflict of interest.

Sample Availability: Samples of the compounds are not available from the authors.

References

1. Li, H. Research on inventory management status of Chinese dairy production enterprises. *China Manag. Inf.* **2021**, *24*, 42–43. (In Chinese)
2. Bai, Y.; Li, L.; Wang, F.; Zhang, L.; Xiong, L. Impact of dairy imports on raw milk production technology progress in China. *Int. J. Environ. Res. Public Health.* **2022**, *19*, 2911. [CrossRef] [PubMed]
3. Chen, R.; Ji, L.; Zhang, S.; Ma, Y. Current situation of China's dairy trade from the perspective of the whole industrial chain. *China Dairy Ind.* **2022**, *02*, 2–7. (In Chinese) [CrossRef]
4. Zhang, X.; Qiao, C.; Fu, S.; Jiao, Y.; Liu, Y. DNA-based qualitative and quantitative identification of bovine whey powder in goat dairy products. *J. Dairy Sci.* **2022**, *105*, 4749. [CrossRef]
5. Mafra, I.; Honrado, M.; Amaral, J.S. Animal species authentication in dairy products. *Foods.* **2022**, *11*, 1124. [CrossRef]
6. Giglioti, R.; Polli, H.; Azevedo, B.T.; Katiki, L.M.; Filho, A.E.V. Detection and quantification of adulteration in milk and dairy products: A novel and sensitive qPCR-based method. *Food Chem. Mol. Sci.* **2022**, *4*, 100074. [CrossRef]
7. Sidra-tul-Muntaha; Iqbal, R.; Yasmin, I.; Tehseen, S.; Khaliq, A.; Chughtai, M.F.J.; Ahsan, S.; Khan, W.A.; Nadeem, M.; Hleba, L.; et al. Safety Assessment of milk and indigenous milk products from different areas of Faisalabad. *J. Microbiol. Biotechnol. Food Sci.* **2020**, *9*, 1197–1203. [CrossRef]
8. Shari, R.V.; Aimei, Y. Media manipulation in the Sanlu milk contamination crisis. *Public Relat. Rev.* **2012**, *38*, 935–937. [CrossRef]
9. Sun, W. Method for Separation and Identification of Bovine Milk Protein and Soy Protein. Master's Thesis, Shaanxi University of Science and Technology, Xi'an, China, 2015. (In Chinese).
10. Ji, Y.; Du, B.; Zhi, X.J. Rapid identification of milk brands and adulteration identification by NIR technology. *Food Res. Dev.* **2016**, *37*, 178–181. (In Chinese)
11. Guo, Z.; Guo, C.; Sun, L.; Zuo, M.; Chen, Q.; El-Seedi, H.R.; Zou, X. Identification of the apple spoilage causative fungi and prediction of the spoilage degree using electronic nose. *J. Food Process Eng.* **2021**, *44*, 8–16. [CrossRef]
12. Xue, H.; Hu, W.; Song, H.; Han, Y.; Yang, Z.Y. Quantitative detection of bovine milk in goat milk by indirect ELISA. *Food Sci.* **2010**, *31*, 370–373. (In Chinese)

13. Abedini, R.; Jahed Khaniki, G.; Molaee Aghaee, E.; Sadighara, P.; Nazmara, S.; Akbari-Adergani, B.; Naderi, M. Determination of melamine contamination in chocolates containing powdered milk by high-performance liquid chromatography (HPLC). *J. Environ. Health Sci. Eng.* **2021**, *19*, 165–171. [CrossRef]
14. Baptista, M.; Cunha, J.T.; Domingues, L. DNA-based approaches for dairy products authentication: A review and perspectives. *Trends Food Sci. Technol.* **2021**, *109*, 386–397. [CrossRef]
15. Zhang, Y.; Meng, Z.; Qiu, K.; Dong, S.; Wu, Z.; Guo, X.; Zhong, Q. Species-specific PCR method to identify animal-derived ingredients of pork, beef, mutton, chicken and duck in canned food. *Food Ferment. Ind.* **2021**, *47*, 164–169. (In Chinese) [CrossRef]
16. Chen, Y.; Zhang, J.; Ge, Y. A bibliometric analysis on technology innovation of food authentication. *J. Food Saf. Qual.* **2019**, *10*, 8183–8194. (In Chinese) [CrossRef]
17. Bi, S.; Yue, S.; Zhang, S. Hybridization chain reaction: A versatile molecular tool for biosensing, bioimaging, and biomedicine. *Chem. Soc. Rev.* **2017**, *46*, 4281. [CrossRef]
18. Saiki, R.K.; Scharf, S.; Faloona, F.; Mullis, K.B.; Horn, G.T.; Erlich, H.A.; Arnheim, N. Enzymatic amplification of β-globin genomic sequences and restrictions site analysis for diagnosis of sickle cell anaemia. *Science* **1985**, *230*, 1350–1354. [CrossRef]
19. Saiki, R.K.; Gelfand, D.; Stoffel, S.; Scharf, S.; Higuchi, R.; Horn, G.; Mullis, K.; Erlich, H. Primer-Directed Enzymatic Amplification of DNA with a Thermostable DNA Polymerase. *Science* **1988**, *239*, 487–491. [CrossRef]
20. Barany, F. Genetic disease detection and DNA amplification using cloned thermostable ligase. *Proc. Nat. Acad. Sci. USA* **1991**, *88*, 189–193. [CrossRef]
21. Compton, J. Nucleic acid sequence-based amplification. *Nature* **1991**, *350*, 91–92. [CrossRef]
22. Fahy, E.; Kwoh, D.Y.; Gingeras, T.R. Self-sustained sequence replication (3SR): An isothermal transcription-based amplification system alternative to PCR. *Genome Res.* **1991**, *1*, 25–33. [CrossRef] [PubMed]
23. Vincent, M.; Xu, Y.; Kong, H. Helicase-dependent isothermal DNA amplification. *Embo. Rep.* **2004**, *5*, 795–800. [CrossRef] [PubMed]
24. Reid, M.S.; Le, X.C.; Zhang, H. Exponential isothermal amplification of nucleic acids and assays for proteins, cells, small molecules, and enzyme activities. *Angew. Chem. Int. Ed.* **2018**, *57*, 11856–11866. [CrossRef]
25. Zhang, D.Y.; Seelig, G. Dynamic DNA nanotechnology using strand-displacement reactions. *Nature Chem.* **2011**, *3*, 103–113. [CrossRef] [PubMed]
26. Lutz, S.; Weber, P.; Focke, M.; Faltin, B.; Hoffmann, J.; Müller, C.; Mark, D.; Roth, G.; Munday, P.; Armes, N. Microfluidic lab-on-a-foil for nucleic acid analysis based on isothermal recombinase polymerase amplification (RPA). *Lab. Chip.* **2010**, *10*, 887–893. [CrossRef]
27. Yue, S.; Li, Y.; Qiao, Z.; Song, W.; Bi, S. Rolling circle replication for biosensing, bioimaging and biomedicine. *Trends Biotechnol.* **2021**, *39*, 1160–1172. [CrossRef]
28. Notomi, T.; Okayama, H.; Masubuchi, H.; Yonekawa, T.; Watanabe, K.; Amino, N.; Hase, T. Loop-mediated isothermal amplification of DNA. *Nucl. Acids Res.* **2000**, *28*, e63. [CrossRef]
29. Bodulev, O.L.; Sakharov, I.Y. Isothermal nucleic acid amplification techniques and their use in bioanalysis. *Biochemistry* **2020**, *85*, 147–166. [CrossRef]
30. Wang, D.; Brewster, J.; Paul, M.; Tomasula, P. Two methods for increased specificity and sensitivity in loop-mediated isothermal amplification. *Molecules* **2015**, *20*, 6048–6059. [CrossRef]
31. Wang, D.; Wang, Y.; Wang, A. Study on false-positive amplification of loop-mediated isothermal amplification. *J. Xuchang Univ.* **2015**, *34*, 81–83. (In Chinese)
32. Wang, D.; Wang, Y.; Zhang, M.; Zhang, Y.; Sun, J.; Song, C.; Xiao, F.; Ping, Y.; Pan, C.; Hu, Y.; et al. Ladder-shape melting temperature isothermal amplification of nucleic acids. *BioTechniques* **2021**, *71*, 359–369. [CrossRef]
33. Kong, A.; Ai, X.; Wang, D. Study on the feasibility of LMTIA technology for detecting HPV nucleic acid. *J. Xuchang Univ.* **2022**, *41*, 72–75. (In Chinese)
34. Gu, M.; Xiao, F.; Wang, D.; Ding, C. Research progress of nucleic acid detection technology for meat provenance detection. *J. Food Saf. Qual.* **2021**, *12*, 7514–7519. (In Chinese) [CrossRef]
35. Ma, C.; Jing, H.; Zhang, P.; Han, L.; Zhang, M.; Wang, F.; Niu, S.; Shi, C. Ultrafast and one-step assay for visual detection of RNA virus. *Chem. Commun.* **2018**, *54*, 3118–3121. [CrossRef]
36. Zhu, K.; Kang, H.; Wang, D. Detection of pork components in Common meat products by visual LAMP. *Food Sci.* **2019**, *40*, 296–302. (In Chinese)
37. Zhang, Y.; Wang, Y.; Ouyang, X.; Wang, D.; Xiao, F.; Sun, J. Development of a Ladder-shape melting temperature isothermal amplification (LMTIA) assay for the identification of cassava component in sweet potato starch noodles. *Molecules* **2022**, *27*, 3414. [CrossRef]
38. Chen, G.; Chen, R.; Ding, S.; Li, M.; Wang, J.; Zou, J.; Du, F.; Dong, J.; Cui, X.; Huang, X.; et al. Recombinase assisted loop-mediated isothermal DNA amplification. *Analyst* **2019**, *145*, 440–444. [CrossRef]
39. Ding, S.; Chen, G.; Wei, Y.; Dong, J.; Du, F.; Cui, X.; Huang, X.; Tang, Z. Sequence-specific and multiplex detection of COVID-19 virus (SARS-CoV-2) using proofreading enzyme-mediated probe cleavage coupled with isothermal amplification. *Biosens. Bioelectron.* **2021**, *178*, 113041. [CrossRef]
40. Wang, Y.; Wang, B.; Wang, D. Development of a Ladder-shape melting temperature isothermal amplification assay for detection of duck adulteration in beef. *J. Food Prot.* **2022**, *85*, 1203–1209. [CrossRef]

41. Wang, B.; Shang, J.; Xu, D.; Wang, D. Detection of mutton component by Ladder-shape melting temperature isothermal amplification (LMTIA) Method. *J. Xuchang Univ.* **2022**, *41*, 58–61. (In Chinese)
42. Khatun, M.A.; Hossain, A.; Hossain, M.S.; Munshi, M.K.; Huque, R. Detection of species adulteration in meat products and Mozzarella-type cheeses using duplex PCR of mitochondrial cytb gene: A food safety concern in Bangladesh. *Food Chem. Mol. Sci.* **2021**, *2*, 100017. [CrossRef] [PubMed]

molecules

MDPI

Article

Elemental Fingerprinting of Pecorino Romano and Pecorino Sardo PDO: Characterization, Authentication and Nutritional Value

Andrea Mara [1,*], Marco Caredda [2], Margherita Addis [2], Francesco Sanna [3], Mario Deroma [4], Constantinos A. Georgiou [5,6], Ilaria Langasco [1], Maria I. Pilo [1], Nadia Spano [1] and Gavino Sanna [1,*]

1 Department of Chemical, Physical, Mathematical and Natural Sciences, University of Sassari, Via Vienna 2, I-07100 Sassari, Italy; ilangasco@uniss.it (I.L.); mpilo@uniss.it (M.I.P.); nspano@uniss.it (N.S.)
2 Department of Animal Science, Agris Sardegna, S.S. 291 Sassari-Fertilia, Km. 18,600, I-07040 Sassari, Italy; mcaredda@agrisricerca.it (M.C.); maddis@agrisricerca.it (M.A.)
3 Department of Environmental Studies, Crop Protection and Production Quality Agris Sardegna, Viale Trieste 111, I-09123 Cagliari, Italy; fsanna@agrisricerca.it
4 Department of Agriculture, University of Sassari, Viale Italia, 39A, I-07100 Sassari, Italy; mderoma@uniss.it
5 Chemistry Laboratory, Department of Food Science and Human Nutrition, Agricultural University of Athens, 75 Iera Odos, 118 55 Athens, Greece; cag@aua.gr
6 FoodOmics.GR Research Infrastructure, Agricultural University of Athens, 118 55 Athens, Greece
* Correspondence: a.mara@studenti.uniss.it (A.M.); sanna@uniss.it (G.S.); Tel.: +39-079229500 (G.S.)

Abstract: Sardinia, located in Italy, is a significant producer of Protected Designation of Origin (PDO) sheep cheeses. In response to the growing demand for high-quality, safe, and traceable food products, the elemental fingerprints of Pecorino Romano PDO and Pecorino Sardo PDO were determined on 200 samples of cheese using validated, inductively coupled plasma methods. The aim of this study was to collect data for food authentication studies, evaluate nutritional and safety aspects, and verify the influence of cheesemaking technology and seasonality on elemental fingerprints. According to European regulations, one 100 g serving of both cheeses provides over 30% of the recommended dietary allowance for calcium, sodium, zinc, selenium, and phosphorus, and over 15% of the recommended dietary intake for copper and magnesium. Toxic elements, such as Cd, As, Hg, and Pb, were frequently not quantified or measured at concentrations of toxicological interest. Linear discriminant analysis was used to discriminate between the two types of pecorino cheese with an accuracy of over 95%. The cheese-making process affects the elemental fingerprint, which can be used for authentication purposes. Seasonal variations in several elements have been observed and discussed.

Keywords: pecorino cheese; elemental fingerprint; authentication; food quality; food safety; ICP-MS; ICP-OES; chemometrics

Citation: Mara, A.; Caredda, M.; Addis, M.; Sanna, F.; Deroma, M.; Georgiou, C.A.; Langasco, I.; Pilo, M.I.; Spano, N.; Sanna, G. Elemental Fingerprinting of Pecorino Romano and Pecorino Sardo PDO: Characterization, Authentication and Nutritional Value. *Molecules* **2024**, *29*, 869. https://doi.org/10.3390/molecules29040869

Academic Editor: Jeongmi Lee

Received: 15 January 2024
Revised: 6 February 2024
Accepted: 8 February 2024
Published: 16 February 2024

1. Introduction

Dairy products and milk are among the most valuable foods due to their high nutritional value. Milk is the primary food of mammals at birth and contains, on average 4.9% of lactose, 6.2% of proteins, 7.9% of fats, 1.93 g kg^{-1} of calcium, 1.58 g kg^{-1} of phosporus and vitamins (vitamin A, 146 IU; vitamin D, 0.18 IU) [1]. Dairy products are derived from the milk of major and minor ruminant species, such as cows, buffaloes, sheep, and goats. In addition to their nutritional properties, dairy products are important for the economy and traditions of many countries. Among them, Italy is one of the most recognized in the world to produce protected designation of origin (PDO) products. The most widely renowned dairy products include Parmigiano Reggiano, Grana Padano, and Pecorino Romano.

Pecorino Romano (PR) is a sheep cheese primarily produced in Sardinia, an Italian region where the sheep dairy industry is economically relevant [2]. In fact, two other PDO sheep's milk cheeses are produced here: Pecorino Sardo PDO (PS) and Fiore Sardo PDO. The production of PDO cheeses from sheep's milk is the main source of income for the livestock industry on the island. Semi-extensive farming is the primary method for rearing milk sheep in Sardinia. This results in cheeses that exhibit unique sensorial properties due to the distinguishing features of pastures and climate.

Despite its relevance to the industry, the economic model is fragile due to its dependence on price fluctuations of PR. This often results in farms failing to cover production costs during times of price drops [2]. For these reasons, recent studies have aimed to enhance economic performance and sustainability of the supply chain [3], develop new marketing strategies, improve farm technologies [4], and diversify the production [5,6]. This latter strategy appears to be the most promising in reducing the reliance of milk costs on PR. Another key method for enhancing product value is to capitalize on consumer awareness of dairy quality and nutritional properties [7,8]. For instance, vitamins and minerals are of significant interest because of their association with various health benefits [7,9].

Elements such as Na, Mg, K, Ca, Fe, Cu, Zn and Se are essential for supporting the immune response, cellular processes, and antioxidant defenses [10,11]. These elements can be obtained naturally through a balanced and tailored diet, which prevents health complications caused by deficiency or excess. On the other hand, toxic elements such as As, Cd, Hg and Pb pose health risks to humans at any concentration [12]. Anthropogenic activities often lead to pollution by toxic elements, which contaminate food through water, soil, and air [13]. To ensure food safety, regulations and safety measures have been implemented to monitor and limit the presence of toxic elements in food. For instance, the European community has established maximum levels of toxic elements in food [14,15]. However, most of the toxic elements in milk and milk products are not currently regulated, such as As, Hg and Cd, while the limit for Pb in such matrices is 0.020 mg kg^{-1} [14]. Previous studies have investigated the dietary intake of toxic metals from milk and its derivatives [16–18].

In addition to nutritional and safety aspect, the elemental composition of foods [19] can provide valuable information for authentication [20], traceability [21], and origin assessment of dairy products [22]. The concentration of elements in foods can be affected by various factors, including climate and translocation from soil, water, and air [19]. Research has shown that elemental fingerprints of dairy products can be used to discriminate their geographical origin [23–25], verify their PDO authenticity [20,26–29], identify breeding methods [30], assess production processes [31], and trace the production chain [32]. Accurate sampling is necessary to achieve these goals and encompass all variables that can influence the elemental fingerprint. This includes seasonality, product processing steps, soil characteristics, pollution sources, and class variability [33].

For several decades, this research group has concentrated on the valorization [34–38], quality protection [39–42], classification [43], and food safety [44–46] of dairy products from Sardinia. Therefore, the potential of elemental analysis in food valorization and authentication was evaluated in this study by measuring the elemental fingerprint of two Sardinian PDO sheep cheeses. The concentrations of 31 elements in 200 samples of PR and PS were determined using inductively coupled plasma-optical emission spectroscopy (ICP-OES) and inductively coupled plasma-mass spectrometry (ICP-MS). The main objective was to evaluate nutritional properties and food safety to enhance products and record data for food authentication studies, e.g., for geographical discrimination. In addition, the effects of cheesemaking and seasonality on elemental fingerprints were evaluated.

2. Results and Discussion

2.1. Elemental Composition of Pecorino Romano PDO and Pecorino Sardo PDO

The elemental analysis of PR and PS was conducted using ICP-OES to determine the concentrations of macroelements (Ca, K, Mg, Na, P, and S), and ICP-MS to measure the amounts of trace elements (Zn, Fe, Mn, Cu, Se, Rb, Sr, Al, B, Co, Ni, Cr, V, Li, and Ag) and toxic elements (As, Cd, Hg, Pb, Sn, Sb, Tl, Te, Bi, and U). The results are presented in Table 1 and are expressed in kg of dry matter in the cheese.

Table 1. Elemental analysis of Pecorino Sardo PDO (n = 97) and Pecorino Romano PDO (n = 103).

Element		Pecorino Sardo PDO			Pecorino Romano PDO		
		Min	Mean ± st.dev.	Max	Min	Mean ± st.dev.	Max
Macro (mg kg^{-1})	Ca	10,000	14,000 ± 1000	16,000	10,000	14,000 ± 1000	16,000
	K	1000	1300 ± 200	1600	700	1000 ± 100	1300
	Mg	600	700 ± 50	800	500	600 ± 40	800
	Na	5000	8000 ± 1000	11,000	17,000	25,000 ± 5000	38,000
	P	8000	9000 ± 500	10,000	7000	9000 ± 700	10,000
	S	500	700 ± 100	1000	500	1000 ± 200	1400
Trace elements (μg kg^{-1})	Zn	37,000	56,000 ± 9000	78,000	18,000	47,000 ± 7500	60,000
	Fe	2600	7000 ± 3000	14,300	2400	6000 ± 950	8500
	Mn	560	850 ± 100	1250	310	800 ± 100	1150
	Cu	600	1200 ± 500	2100	500	1000 ± 350	1800
	Se	210	340 ± 90	530	220	400 ± 100	580
	Rb	1000	1700 ± 500	2600	600	1600 ± 500	2600
	Sr	7800	13,400 ± 2500	17,900	4400	14,300 ± 2500	20,500
	Al	200	6000 ± 3000	13,900	2500	6000 ± 2000	11,300
	B	<54	2000 ± 2000	7100	<54	8000 ± 8000	30,000
	Co	0.9	4 ± 1	9	1.3	4 ± 1	6
	Ni	<10	30 ± 10	60	<10	27 ± 5	40
	Cr	<3.1	40 ± 20	95	<3.1	20 ± 10	50
	V	5	10 ± 5	21	7	15 ± 5	24
	Li	<55	<55	<55	<55	<55	<55
	Ag	<1.6	5 ± 5	8	<1.6	5 ± 5	10
Toxic elements (μg kg^{-1})	As	<3.3	6 ± 1	8.4	5.7	8 ± 1	10.8
	Cd	0.5	1 ± 0.5	1.7	0.5	1.2 ± 0.5	1.9
	Hg	<30	<30	<30	<30	<30	<30
	Pb	<3.4	20 ± 10	40	<3.4	20 ± 10	45
	Sn	<2.4	20 ± 10	54	<2.4	10 ± 10	32
	Sb	<3.6	12 ± 5	16	<3.6	10 ± 5	18
	Tl	<0.5	1.9 ± 0.5	2.3	<0.5	<0.5	<0.5
	Te	<1.2	130 ± 50	220	<1.2	9 ± 5	15
	Bi	<0.5	<0.5	<0.5	<0.5	2 ± 1	2.8
	U	<0.19	1 ± 1	5.2	<0.19	2 ± 1	7.1

Both cheeses had comparable levels of Ca, Mg, K and P in terms of macroelements, with an order of abundance that reflected their concentration in the original milk: Ca > P > K > S ≥ Mg [1,9]. However, due to the distinct salting process employed, the Na concentration in PR was higher than that in PS. Typically, the NaCl concentration in PR ranges from 3% to 7%, whereas in PS it rarely exceeds 2%. Even in terms of trace elements, both cheeses have a similar elemental content. Consistent with the initial milk composition, the trace elements found in highest abundance were Zn, Fe, and Cu. Both PR and PS contained similar amounts of Se, Rb, Sr, and Al. Other trace elements, such as Co, Ni, Cr, V, Li, and Ag, were present in both cheeses at levels near or below the limit of quantification.

Regarding toxic elements, both cheeses contained low levels of As, Cd and Pb. Hg was never detected. Other toxic elements, such as Sn, Sb, Te, Tl, Bi, and U, were generally either not quantified or present at very low levels. It is worth noting that the level of Te in PS was significantly higher than that in PR. However, the European Food Safety Agency (EFSA) is currently investigating the potential toxicity of Te [47].

Finally, a semi-quantitative analysis of rare earth elements (REEs) was preliminarily performed. The REEs were seldom detected above the instrumental detection limit, with a few exceptions for the LREEs. Further investigations will be conducted to optimize the limits of quantification of the analytical method and enable the determination of markers for traceability of the production chain [32].

To the best of our knowledge, the determination of trace elements in PR and/or PS has rarely been accomplished. Previous literature has mainly focused on quantifying macroelements [28,48,49], with only occasional attention given to trace elements such as Zn, Fe, Se, Cu [48], Ba [28], and Al, Ba, Cd, Co, Cr, Cu, Fe, Mn, Ni, Pb, Pt, Sr, and Zn [49]. The data obtained in this study are in good agreement with those of previous studies. Table S1 enables a comparison of the elemental compositions of PS and PR as measured in this study and in the literature.

2.2. Differentiation Due to Cheese-Making Process Technology

Elemental fingerprinting has been reported in the literature as a method for authenticating cheeses [23,24,26–28,50]. This technique is commonly used to discriminate cheeses made from milk of different animal origins [26,27,50], from different geographical areas [23,28], or from significantly different cheese-making processes, such as final moisture content and salting [24,26]. Samples were collected from various dairies in Sardinia (Italy) and varied in cheese-making technologies and production period (seasonality).

Principal Component Analysis (PCA) was used for data visualization. The data was cleaned by removing any elements that were not quantified in at least 90% of the samples or were not significant for the analysis. Additionally, Na was excluded as a variable to eliminate the influence of the salting process. Outliers were identified and removed using T^2 and Q statistics after performing a preliminary PCA with $p > 0.05$. The results of the PCA are presented in Figure 1.

PC1 and PC2 accounted for 24.5% and 19.7% of the total variance, respectively. Figure 1A shows that the most abundant trace elements, such as Zn and Cu, are characterized by positive values of PC1, while macroelements, including Ca, Mg, P, and K, are characterized by positive PC2 values. Looking at the score plot (Figure 1B), positive PC1 values tended to occur in the PS cluster (red samples), which was associated with a higher concentration of trace elements, while the PR cluster (black samples) tended to have negative PC1 values. The differentiation between the two clusters was accentuated upon observing PC3, which explained 12.1% of the variance (3D score plot, Figure 1C). This evidence suggests that the elemental fingerprint may be used to discriminate between the two types of cheese. Linear Discriminant Analysis (LDA) was used for classification. The MANOVA test showed a significant difference between the two groups (F (15, 180), Wilks = 0.175, approx. $F = 56.78$, $p < 0.001$). Prior to LDA, the dataset was randomized and split into a training set (n = 140) and a test set (n = 55). The results obtained from cross-validation and prediction are reported in Table 2.

Table 2. Results of the LDA performed for the discrimination based on cheese type. Confusion matrix and accuracy in cross-validation (training) and prediction (testing).

Confusion Matrix					
Training	Romano	Sardo	Testing	Romano	Sardo
Romano	67	3	Romano	32	0
Sardo	1	69	Sardo	2	21
Accuracy					
Romano	Sardo	Total	Romano	Sardo	Total
95.7%	98.6%	97.1%	100%	91.3%	95.7%

The levels of discrimination achieved in cross-validation (97.1%) and prediction (95.7%) were highly accurate. The elemental fingerprint can discriminate between PR and PS using macro-elements (i.e., Ca, K, Mg, P, and S) and trace elements (i.e., Zn, Fe, Mn, Cu, Se, Rb, Sr, Al, Co, and V). These findings were consistent when LDA was used to analyze data from samples produced by three farms that yielded both PR and PS during the same period. The statistical significance of the data was reduced due to the smaller sample size (n = 62).

However, the Principal Component Analysis (PCA) in Figure S1 showed that the samples were distinguishable, and the LDA successfully classified them with an accuracy of 98.1% in cross-validation.

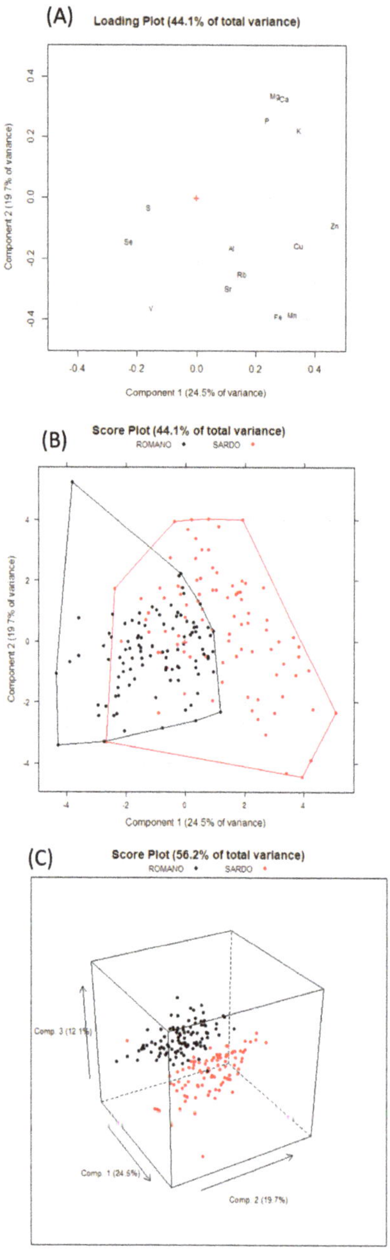

Figure 1. Principal component analysis performed on data obtained from the elemental determination of 14 elements on 196 pecorino samples: (**A**) loading plot; (**B**) score plot; (**C**) 3D score plot. Object colored according to cheese type.

Among the literature reviewed, the study by Di Donato et al. [28] was the most comparable to the present study as it adopted a similar approach for authenticating Italian Pecorino cheese. However, it is important to note that the samples in their study were geographically diverse, collected from three different regions of Italy. In contrast, the differences found in our study were solely attributed to the cheese production method used.

The study confirmed the impact of cheese-making technology on the elemental fingerprint. The differences in the chemical equilibrium of milk's various elements, as well as the chemical form (ionic soluble or colloidal bound to milk proteins and fats), and the variations in cheese-making techniques (coagulation, wheying, and salting) could explain the distinct elemental fingerprints of the two cheeses analyzed in this research. During cheese production, the main stages that lead to a significant variance in the concentration of elements found in the cheese are salting and coagulation.

Salting triggers an osmotic phenomena, resulting in fluctuations in the levels of unbound minerals, which can cause a loss of water and cationic elements such as Na, K, Al, Cd, Co, and Rb. On the other hand, the elements bound to caseins and fats, such as Ca, P, Fe, Mg, Mn, Ni, Pt, and Zn, become concentrated.

During coagulation, there is a non-uniform distribution of elements between the curd and whey. Na and K are soluble elements and tend to be distributed in the aqueous phase (whey). On the other hand, Ca, Mg and P are associated in different proportions with the colloidal suspension of casein micelles and are more concentrated in the curd during cheese production [51]. Several studies have shown that the soluble phase of sheep's milk may contain different percentages of Ca, Mg and P, with deviations from the total content ranging from 20–25%, 35–64% and 35–40%, respectively [52–54]. Currently, there is limited information available on the distribution of trace elements during sheep milk cheese production. However, data indicate that Zn and Mn are primarily distributed in the curd, accounting for about 90%, while Fe and Cu account for about 70% [52]. Fluctuations in pH, temperature, and milk storage conditions affect the equilibrium between soluble and colloidal forms. Generally, a decrease in pH and temperature shifts the balance towards soluble ionic forms, while an increase in pH and temperature favors solubilization. This results in an increase in the retention of certain elements in the curd during cheese production.

Based on these considerations, the various conditions of acidification (which are more intense in PR compared to PS), curd breaking (which is more extensive in PR compared to PS), cooking (with curd cooking at 45 °C in PR and 43 °C in PS), whey removal, heating, curd cooling, and salting may have induced differences in the balance between curd and whey. As a result, the different cheesemaking technologies favor the retention of certain elements, especially Zn, Fe, Cu, and Mn in PS compared to PR (see Figure 1). Milder acidification and breaking of the curd may have resulted in less demineralization during the production of PS. This could have led to greater retention of trace elements that are partially bound to casein micelles, such as Zn, Mg, Fe, and Cu, as previously observed. Additionally, rapid lactic fermentation, followed by effective whey removal, promotes curd demineralization [55].

2.3. *Effect of Seasonality*

In Sardinian sheep farming births are synchronized and the lactation period starts in November and ends in June–July. The chemical composition of sheep's milk changes during this period, depending on the diet, lactation stage, and climate [4]. The concentration of minerals and major classes of compounds can be affected by the lactation stage [56]. Therefore, we evaluated the effect of seasonality on the elemental composition of PR and PS cheeses using PCA and ANOVA.

The PCA analysis in Figure S2 shows a distinct trend for Pecorino Romano PDO cheese. The loading plot indicates that PC1 describes the correlation between trace element concentrations and seasonality. Negative values indicate elements that are more abundant in PR produced in summer (V, Al, Rb, and Fe), while positive values indicate elements that are more abundant in PR produced in winter (Zn and Cu). The ANOVA results confirmed that seasonality had an impact on 12 out of 15 elements (see Figure S3). Winter-produced cheeses had the highest concentrations of Zn, Cu, K, and Mn, while spring-produced cheeses had the highest concentrations of Ca, Mg, P, and S. Summer-produced cheeses had the highest concentrations of Rb, Fe, Al, and V. The concentrations of Se, Na, and Sr were not affected by seasonal variations.

In contrast, the impact of seasonality on Pecorino Sardo PDO is relatively minor. Although the PCA did not reveal any clear trends (see Figure S4), the ANOVA indicates a significant effect of seasonality on 8 out of 15 elements (Zn, Ca, K, P, S, Cu, Rb, and V), as shown in Figure S5. Notably, the trends for these elements were like those observed for PR, with higher concentrations of Ca, P, and Mg found in PS during the spring season. Additionally, the concentration of Cu was highest in winter cheese.

As expected, both types of pecorinos exhibit similar seasonal variations in their composition, reflecting the composition of sheep's milk. Ca is closely linked with P in casein micelles, which provide the structure and stability of the micelles. Colloidal calcium phosphate links the casein submicelles together, occupying 6% of the micellar structure. Therefore, there is a positive correlation between Ca, P, and casein content in ruminant milk [57]. Both cheeses (PS and PR) exhibit the highest concentrations of Ca, Mg, P, and S during the spring season when the amount of casein in the ewe's milk reaches its daily maximum level [4]. S is not directly involved in the stabilization of micelles, but it is present in proteins, specifically whey proteins (cysteine and cystine amino acids). Therefore, the higher concentration of sulfur in spring cheeses could be linked to the protein concentration found in the milk of spring sheep. Despite this, the cheese still contains only small amounts of whey proteins. When examining the alkaline elements, no evident trends were found. Na cannot be evaluated due to the salting process. K remained constant in winter and spring but decreased significantly in summer. Rb in PR increased from winter to summer, but this trend was not observed in PS. This difference was likely due to different production methods. Regarding trace elements, there were two opposite trends observed. At the start of the lactation period, the concentrations of Zn, Mn, and Cu were highest. However, towards the end of the lactation period, the concentrations of Fe and V tended to increase.

The elemental composition of sheep's milk during lactation is subject to seasonal variations, which are influenced by various factors such as the lactation stage, nutritional status of the animal, as well as environmental and genetic factors [51,58]. The mineral content in milk is weakly affected by ruminant feeding because the maternal skeleton tends to demineralize during periods when dietary mineral intake does not meet the mineral requirements of the newborn, thus, compensating for the deficit [59]. Skeletal demineralization typically occurs during periods of high mineral demand, such as early lactation and colostrum production [60]. The influence of the lactation stage on the mineral composition of milk is not well-documented. In bovine milk, Ca, P, Mg, and Na levels tend to increase towards the end of the lactation period [61]. This is likely due to increased permeability of the mammary epithelium as lactation progresses [62]. Finally, the mineral content of milk can also be influenced by the animal's health status and genetic type. The concentration of most minerals in milk decreases when mastitis is present in the mammary gland, except for sodium and chloride ions, which increase instead [63].

2.4. Nutritional and Safety Aspects

Milk and dairy products are considered highly nutritious. Mineral content is an important factor in determining food value, according to consumer preferences. Cheese is a well-known source of minerals, especially Ca, P, and Mg. Casein peptides in milk or cheese prevent the precipitation of calcium in the intestine, making it easily bioavailable [64]. Although the etiology of osteoporosis is complex, adequate calcium intake during childhood and adolescence is important for developing high peak bone mass. Maximizing bone mass early in life is considered a crucial preventive factor against osteoporosis [65]. This study's results indicate that Pecorino Romano PDO and Pecorino Sardo PDO are sources of several nutritional elements. Figure 2 displays the daily mineral intake for both cheeses across genders. The data were calculated based on the Dietary Recommended Intakes (DRI) published by the US Department of Health and Human Services, National Institutes of Health [66].

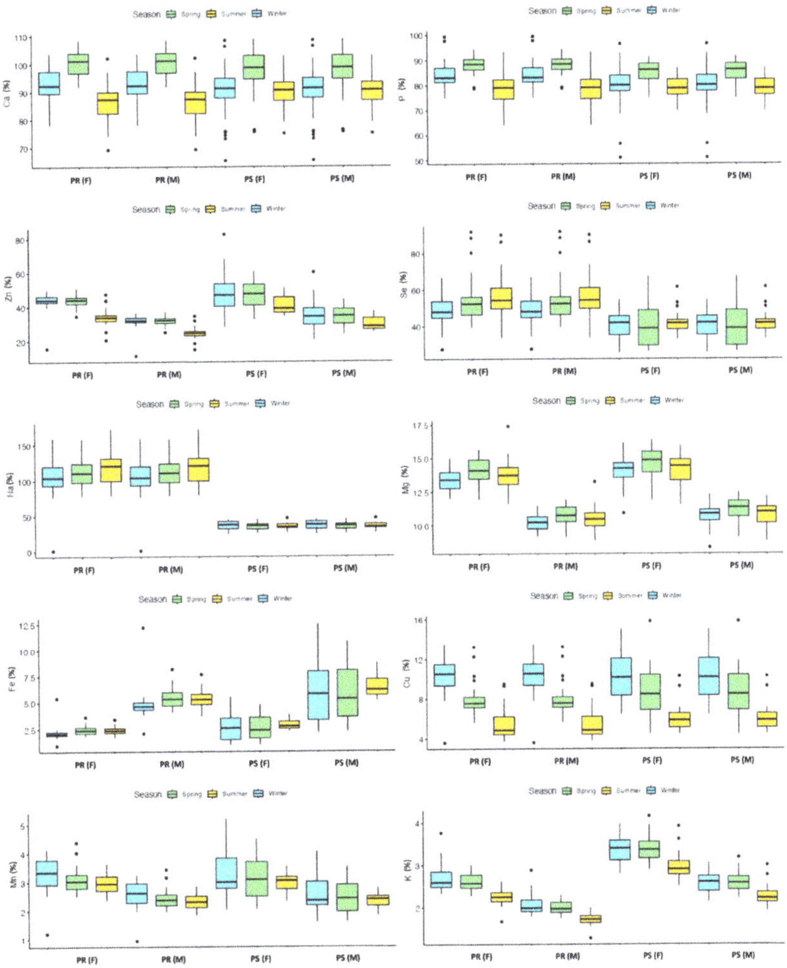

Figure 2. Mineral daily intakes for adult males and females (17–70 years old) for the consumption of 100 g of Pecorino Romano PDO and Pecorino Sardo PDO.

Figure 2 shows that the daily consumption of both cheeses meets the Recommended Dietary Allowances (RDAs) and Adequate Intakes (AIs) for many elements (daily portion of 100 g). According to European guidelines [67,68], both cheeses are rich in Ca, P, Zn, and Se (DRI > 30%). Additionally, PR and PS cheeses are potential sources of Mg (females, DRI > 15%) and Cu (DRI > 15%). The Cu content in cheeses may vary seasonally, as shown in Figures S3–S5. Therefore, it may be possible to produce cheeses with high Cu content during the winter season. These findings are significant as they allow dairies and protection Consortia to implement nutritional labeling in accordance with European regulations [69]. As for toxic elements such as As, Cd, Hg, Pb, and Sn, both PR and PS cheeses have a high level of food safety. Additionally, the concentrations of Tl, Bi, and U were frequently below their limits of quantification. Moreover, the cheeses analyzed in this study were obtained from various locations (see Figure 3), indicating the high level of food safety in Sardinian sheep's milk production.

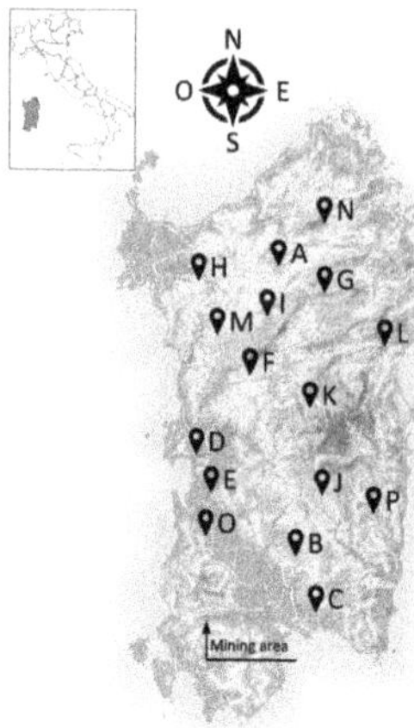

Dairy Factory	Pecorino Romano			Pecorino Sardo			Total
	Winter	Spring	Summer	Winter	Spring	Summer	
A	2	2	3				7
B	3	3	3				9
C				6	6	4	16
D	3	3	3	6	5	2	22
E	3	3	3				9
F	3	3	3				9
G	2	2	3				7
H	2	3	3				8
I	3	3	3				9
J				6	6	4	16
K	3	3	3				9
L	2	4	3				9
M	3	3	3	6	4	2	21
N	3	3	3	6	4	2	21
O				6	4	2	12
P				6	6	4	16
Total	32	35	36	42	35	20	200

Figure 3. Description of pecorino sampling in terms of dairies, samples, and period of production.

3. Materials and Methods

3.1. Samples

A total of 200 samples of Pecorino cheese produced in Sardinia in 2021 were obtained from 16 dairy farms that collected milk from livestock farming in surrounding areas. Two PDO sheep cheeses were considered: PR (n = 103) and PS (n = 97). The production technologies adhere closely to the specifications outlined in their respective consortium regulations [70,71]. Pasteurized whole milk is curdled using cultures of milk enzymes from the milk's place of origin. To produce PR, the milk is coagulated at 38–40 °C and the curd is cooked at 45–48 °C. The resulting 20–35 kg wheels mature for 5–18 months. On the other hand, PS is produced by coagulating the milk at 35–39 °C, cooking the curd at 43 °C and ripening the 1.7–4.0 kg wheels for 2–6 months. The salting process can occur through both dry and wet methods. For PR dry salting, the side and plate of the cheese receive about 2800 g of NaCl distributed thrice over a maximum span of 50 days. Alternatively, the wheels are immersed in a dynamic brine (at 12 °C and a NaCl concentration of 20–22%) for 20 days. Similarly, PS can also be salted dry or wet, but generally, it is kept in brine (24% of NaCl) for 10 h kg^{-1} of cheese at a temperature of 10–12 °C.

In this study, cheeses were produced by each dairy in three periods of the year: winter, 37%; spring, 35%; and summer, 28%. Thus, samples differed in seasoning and cheesemaking. Additionally, the cheeses varied in maturation (PR: 5–18 months, PS: 2–6 months). The wheels were first divided according to what reported in previous studies ISO 707:2008 [72]. Then, aliquots obtained were aggregated, homogenized, and stored at a temperature between −18 °C and −24 °C until analysis. Figure 3 shows the selected information on Pecorino cheese samples.

3.2. Instrumentation and Reagents

Elemental analysis was performed on a NexION 350X spectrometer equipped with an S10 autosampler, a glass concentric nebulizer, a glass cyclonic spray chamber, and a kinetic energy discrimination (KED) collision cell, all from Perkin Elmer (Milan, Italy). The most abundant elements were determined using an OPTIMA 7300 DV spectrometer (Perkin Elmer, Waltham, MA, USA) equipped with a GemTip Cross-Flow II nebulizer (Perkin Elmer, Waltham, MA, USA) and an autosampler (SC-2 DX, Elemental Scientific Inc., Omaha, NE, USA). To determine macroelements (Ca, K, Mg, Na, P, and S) cheese samples were previously dried in a drying oven (Memmert, Schwabach, Germany) and then calcined in a muffle furnace (Gelman Instrument, Opera, Italy). To detect trace elements (Zn, Fe, Mn, Cu, Se, Rb, Sr, Al, B, Co, Ni, Cr, V, Li, and Ag) and toxic elements (As, Cd, Hg, Pb, Sn, Sb, Tl, Te, Bi, and U) samples were digested using an ultraWAVE™ microwave single reaction chamber (SCR) system (Milestone, Sorisole, Italy) equipped with a rotor at 15 positions and 15 cm^3 polytetrafluoroethylene (PTFE) vessels. In all procedures, type I water (resistivity > 18 MΩ cm^{-1}) was produced using a MilliQ Plus System (Millipore, Milan, Italy). Hydrochloric acid (37% *w*/*w*), nitric acid (67–69% *w*/*w*, NORMATON® for ultra-trace analysis), hydrogen peroxide (30% *w*/*w*, NORMATON® for ultra-trace metal analysis), syringes (HENKE-JECT®, 20 cm^3), syringe filters (25 mm diameter, 0.22 µm pores, nylon), paper filters (ashless, Grade 40, Whatman®) and metal-free falcon tubes (15 cm^3 and 50 cm^3) were purchased from VWR (Milan, Italy). Periodic table mix 1 (TraceCERT®, 33 elements, 10 mg dm^{-3} in 10% HNO_3), periodic table mix 2 (TraceCERT®, 17 elements, 10 mg dm^{-3} in 5% HCl), periodic table mix 3 (TraceCERT®, 16 elements, 10 mg dm^{-3} in 5% HNO_3), and certified skimmed milk powder ERM-BD151 were obtained from Sigma-Aldrich (St. Louis, MO, USA). Single standard solutions of Sc, Y, Ge, Rh, Ir, Mo, Sb, Sn, Hg, and U (100–1000 mg dm^{-3} in 2–5% HNO_3) were obtained from LabKings (Hilversum, The Netherlands).

3.3. Sample Preparation

Sample preparation for macroelements analysis was performed as previously described [6]. The sample digestion for trace and toxic elements was made using an ultraWAVE™ microwave single reaction chamber (SRC) system (Milestone, Sorisole). With respect to conventional microwave instruments, SRC technology reaches higher temperatures and pressures and can manage higher amounts of samples using lower amounts of reagents [73]. In accordance with previous studies [73,74], nitric acid and hydrogen peroxide were used as oxidizing agents in this study. Approximately 0.450 g of the sample (exactly weighted on the analytical balance) was treated with 1 cm^3 of HNO_3 (67–69%), 2 cm^3 of H_2O_2 (30%), and 4 cm^3 of ultrapure H_2O.

The digestion program is listed in Table 3. After cooling, the samples were collected, diluted to 15 cm^3 using ultrapure H_2O, and filtered using a syringe filter. The final residual acidity determined by titration with 0.1 mol dm^{-3} sodium hydroxide was 2.5 ± 0.2%. To ensure the quality of the analytical data, each digestion batch included a blank and a sample of certified reference material (CRM), ERM-BD151. The same CRM was used to assess the efficiency of microwave acid digestion in terms of matrix effect and trueness.

Table 3. Cheese digestion conditions using an ultraWAVE™ SRC system (Milestone).

Step		Time (min)	Temperature (°C)
1	Heating	25	240
2	Holding	10	240
3	Cooling	ca. 30	<40

Initial pressure: 4 MPa; Release pressure rate: 0.8 MPa min^{-1}; Rotor: 15 positions; Vessels: 15 cm^3 (PTFE), Method: 0.450 g cheese + 1 cm^3 HNO_3 (67–69%) + 2 cm^3 H_2O_2 (30%) + 4 cm^3 H_2O.

3.4. Elemental Analysis, Validation, Quality Control and Assurance

Macroelements (Ca, K, Mg, Na, P, and S) were determined using inductively coupled plasma-optical emission spectroscopy (ICP-OES), whereas trace and toxic elements were analyzed using inductively coupled plasma-mass spectrometry (ICP-MS). The instrumental parameters used for the analysis are reported in Table S2 (ICP-OES) and Table S3 (ICP-MS). Further details regarding the ICP-OES method have been previously reported [6], whereas the ICP-MS method was fully developed and validated in this study. For each PDO cheese, three samples were randomly selected and analyzed using the semi-quantitative TotalQuant® method (Syngistix software v 2.3). This preliminary analysis allowed the assessment of the elements that were always within the instrumental detection limit. Excluding these, the elements of interest for a possible ICP-MS quantification were Ag,

Al, As, B, Bi, Cd, Co, Cr, Cu, Fe, Hg, Li, Mn, Ni, Pb, Rb, Sb, Se, Sn, Sr, Te, Tl, U, V, and Zn. Subsequently, the possible presence of polyatomic interferences in the real matrix was ascertained, and for each element, the most suitable analysis mode (STD mode or KED mode) was determined. Validation was accomplished in terms of limits of detection and quantification, precision, and trueness. The validation parameters are listed in Table S4. The limits of detection (LoD) and quantification (LoQ) were calculated according to Currie [75]. Method repeatability ($CV\%_r$) was assessed by analyzing samples in triplicate within the same analytical session, whereas intermediate precision ($CV\%_{IP}$) was calculated using data obtained from different analytical sessions. Finally, trueness was evaluated by analyzing certified milk CRM ERM BD-151 and spiking tests. In the last case, for each analyte, samples were spiked three times at increasing concentration levels. The trueness measured by analyzing the CRM (Table S5) was between 89 $\pm$ 5% (P) and 114 $\pm$ 5% (Na) for the macroelements (ICP-OES method), whereas that measured for the trace and toxic elements (ICP-MS method) ranged from 92 $\pm$ 5% (Cd) to 110 $\pm$ 5% (Se). Furthermore, the recoveries measured by spiking tests ranged from 86 $\pm$ 1% (Ag) to 149$\pm$ 7% (As). The recovery results (Table S4) show that the determination of more than 80% of the elements was bias-free (criteria: *t*-test, p = 95%). Moderate underestimation was observed for Ag and Bi, and slight overestimation for Sr and V. However, the observed bias are acceptable according to the AOAC guidelines [76]. Finally, for Na and As, a meaningful overestimation bias was observed.

Quantification was performed by external calibration using single- and multi-element standard solutions in 2.5% HNO_3. The calibration was performed according to the expected analyte concentrations. Additionally, measurements were performed in triplicate and the data were blank-corrected. To account for signal fluctuations and matrix effects, Rh (50 $\mu g\ dm^{-3}$) and Ir (1 $\mu g\ dm^{-3}$) were used as internal standards. A 60-s wash with a 2% aqueous solution of HNO_3 was introduced between consecutive samples to prevent memory effects.

3.5. Statistical Analysis

Data analysis was performed using R-Studio (v. 4.3.1) and Chemometric Agile Tool (CAT) [77]. The Shapiro–Wilk test was used to confirm the normal distribution of the data. ANOVA and MANOVA tests were used to compare groups, and Tukey's HSD test was used as a post hoc test. The data set was cleaned by removing all elements that were not quantified in at least 90% of the samples or were not significant for the analysis. Elements retained for chemometric analysis were Mg, Ca, P, K, Zn, Cu, Fe, Mg, Al, Rb, Sr, S, Se, and V. Principal Component Analysis (PCA) was performed for data visualization. Linear Discriminant Analysis (LDA) was used to discriminate samples from different categories. Statistical significance was set at $\alpha < 0.05$.

4. Conclusions

Pecorino Romano PDO and Pecorino Sardo PDO are two popular and highly regarded sheep's cheeses. This study aimed to address the lack of knowledge regarding their elemental composition and trace element content. To achieve this, 31 elements (macro, trace, and toxic) were analyzed in a comprehensive sample of cheeses using ICP-based methods. The results showed that these cheeses are rich in essential minerals such as calcium, phosphorus, zinc, and selenium, and could potentially be a source of copper and magnesium.

Additionally, the data allowed for an assessment of the impact of cheesemaking and seasonality on the elemental composition. Linear discriminant analysis showed that the elemental fingerprint can effectively distinguish dairy products based on their production method. This result could be a useful methodology for detecting food fraud involving other cheaper pecorinos.

Supplementary Materials: The following supporting information can be downloaded at: https://www.mdpi.com/article/10.3390/molecules29040869/s1, Figure S1: Principal component analysis performed on Pecorino Sardo and Pecorino Romano produced by 3 farms in the same period: (a) loading plot; (b) score plot.; Figure S2. Principal component analysis performed on Pecorino Romano samples and 14 elements: (a) loading plot; (b) score plot. Object colored according to seasonality; Figure S3: ANOVA analysis of macro and trace elements in Pecorino Romano PDO as a function of the seasonality; Figure S4: Principal component analysis performed on Pecorino Sardo samples and 14 elements: (a) loading plot; (b) score plot. Object colored according to seasonality; Figure S5: ANOVA analysis of macro and trace elements in Pecorino Sardo as a function of the seasonality; Table S1: Average elemental composition of Pecorino cheeses measured in this study and from literature data; Table S2: Instrumental conditions of the ICP-OES OPTIMA 7300 DV, Perkin Elmer; Table S3: Instrumental conditions of the ICP-MS NexION 350X, Perkin Elmer; Table S4: Validation parameters of the ICP-MS method for the elemental analysis of Pecorino cheeses; Table S5: Analysis of the CRM ERM BD-151 (skimmed milk powder).

Author Contributions: Conceptualization, M.A. and G.S.; methodology, A.M., M.A., C.A.G. and G.S.; software, A.M. and M.C.; validation, A.M., M.C. and G.S.; formal analysis, A.M., M.C., M.A., F.S. and G.S.; investigation, A.M., M.D. and I.L.; resources, M.A. and G.S.; data curation, A.M. and M.C.; writing—original draft preparation, A.M., M.C. and G.S.; writing—review and editing, A.M., M.C., M.A., C.A.G., I.L., M.I.P., N.S. and G.S.; visualization, A.M. and M.C.; supervision, M.A., C.A.G. and G.S.; project administration, M.A. and G.S.; funding acquisition, M.A. and G.S. All authors have read and agreed to the published version of the manuscript.

Funding: This research was funded by Sardegna Ricerche, "POR FESR—Azione 1.1.4 Sostegno alle attività collaborative di R&S per lo sviluppo di nuove tecnologie sostenibili, di nuovi prodotti e servizi, Progetti Cluster—La Diversificazione di Prodotto nell'ambito del Pecorino Romano DOP—CUP G77D13000010002.

Institutional Review Board Statement: Not applicable.

Informed Consent Statement: Not applicable.

Data Availability Statement: The data presented in this study are available in article and Supplementary Materials.

Acknowledgments: The authors would like to thank the Consorzio per la Tutela del formaggio Pecorino Romano DOP and the Consorzio per la Tutela del formaggio Pecorino Sardo DOP for their availability and collaboration in the involvement of the dairy companies and in the procurement of cheese samples. The authors greatly thank Demetrios Sotirchos and Georgios Danezis for their support.

Conflicts of Interest: The authors declare no conflicts of interest. The funders had no role in the design of the study; in the collection, analyses, or interpretation of data; in the writing of the manuscript; or in the decision to publish the results.

References

1. Pereira, P.C. Milk Nutritional Composition and Its Role in Human Health. *Nutrition* **2014**, *30*, 619–627. [CrossRef] [PubMed]
2. Pulina, G.; Milán, M.J.; Lavín, M.P.; Theodoridis, A.; Morin, E.; Capote, J.; Thomas, D.L.; Francesconi, A.H.D.; Caja, G. Invited Review: Current Production Trends, Farm Structures, and Economics of the Dairy Sheep and Goat Sectors. *J. Dairy Sci.* **2018**, *101*, 6715–6729. [CrossRef] [PubMed]
3. Vagnoni, E.; Franca, A.; Porqueddu, C.; Duce, P. Environmental Profile of Sardinian Sheep Milk Cheese Supply Chain: A Comparison between Two Contrasting Dairy Systems. *J. Clean. Prod.* **2017**, *165*, 1078–1089. [CrossRef]
4. Pulina, G.; Atzori, A.S.; Dimauro, C.; Ibba, I.; Gaias, G.F.; Correddu, F.; Nudda, A. The Milk Fingerprint of Sardinian Dairy Sheep: Quality and Yield of Milk Used for Pecorino Romano P.D.O. Cheese Production on Population-Based 5-Year Survey. *Ital. J. Anim. Sci.* **2021**, *20*, 171–180. [CrossRef]
5. Camanzi, L.; Arba, E.; Rota, C.; Zanasi, C.; Malorgio, G. A Structural Equation Modeling Analysis of Relational Governance and Economic Performance in Agri-Food Supply Chains: Evidence from the Dairy Sheep Industry in Sardinia (Italy). *Agric. Food Econ.* **2018**, *6*, 4. [CrossRef]
6. Lai, G.; Caboni, P.; Piras, C.; Pes, M.; Sitzia, M.; Addis, M.; Pirisi, A.; Scano, P. Development and Chemico-Physical Characterization of Ovine Milk-Based Ingredients for Infant Formulae. *Appl. Sci.* **2023**, *13*, 653. [CrossRef]

7. Kraus, A. Development of Functional Food with the Participation of the Consumer. Motivators for Consumption of Functional Products. *Int. J. Consum. Stud.* **2015**, *39*, 2–11. [CrossRef]
8. Sajdakowska, M.; Gębski, J.; Guzek, D.; Gutkowska, K.; Żakowska-Biemans, S. Dairy Products Quality from a Consumer Point of View: Study among Polish Adults. *Nutrients* **2020**, *12*, 1503. [CrossRef]
9. de la Fuente, M.A.; Juárez, M. Milk and Dairy Products. In *Handbook of Mineral Elements in Food*; Wiley: Hoboken, NJ, USA, 2015; pp. 645–668, ISBN 9781118654316.
10. Cannas, D.; Loi, E.; Serra, M.; Firinu, D.; Valera, P.; Zavattari, P. Relevance of Essential Trace Elements in Nutrition and Drinking Water for Human Health and Autoimmune Disease Risk. *Nutrients* **2020**, *12*, 2074. [CrossRef]
11. Islam, M.R.; Akash, S.; Jony, M.H.; Alam, M.N.; Nowrin, F.T.; Rahman, M.M.; Rauf, A.; Thiruvengadam, M. Exploring the Potential Function of Trace Elements in Human Health: A Therapeutic Perspective. *Mol. Cell. Biochem.* **2023**, *478*, 2141–2171. [CrossRef]
12. Jaishankar, M.; Tseten, T.; Anbalagan, N.; Mathew, B.B.; Beeregowda, K.N. Toxicity, Mechanism and Health Effects of Some Heavy Metals. *Interdiscip. Toxicol.* **2014**, *7*, 60–72. [CrossRef]
13. de Almeida Ribeiro Carvalho, M.; Botero, W.G.; de Oliveira, L.C. Natural and Anthropogenic Sources of Potentially Toxic Elements to Aquatic Environment: A Systematic Literature Review. *Environ. Sci. Pollut. Res.* **2022**, *29*, 51318–51338. [CrossRef]
14. *EC No 1881/2006*; Setting Maximum Levels for Certain Contaminants in Foodstuffs. European Commission: Luxembourg, 2006; pp. 5–24.
15. *EU No 15/2011*; Amending Regulation (EC) No 2074/2005 as Regards Recognised Testing Methods for Detecting Marine Biotoxins in Live Bivalve Molluscs. European Commission: Luxembourg, 2011; pp. 3–6.
16. Koch, W.; Czop, M.; Iłowiecka, K.; Nawrocka, A.; Wiącek, D. Dietary Intake of Toxic Heavy Metals with Major Groups of Food Products—Results of Analytical Determinations. *Nutrients* **2022**, *14*, 1626. [CrossRef]
17. Crupi, R.; Lo Turco, V.; Gugliandolo, E.; Nava, V.; Potortì, A.G.; Cuzzocrea, S.; Di Bella, G.; Licata, P. Mineral Composition in Delactosed Dairy Products: Quality and Safety Status. *Foods* **2022**, *11*, 139. [CrossRef]
18. Giri, A.; Bharti, V.K.; Kalia, S.; Kumar, B.; Chaurasia, O.P. Health Risk Assessment of Heavy Metals Through Cow Milk Consumption in Trans-Himalayan High-Altitude Region. *Biol. Trace Elem. Res.* **2021**, *199*, 4572–4581. [CrossRef]
19. Zhang, P.; Georgiou, C.A.; Brusic, V. Elemental Metabolomics. *Brief. Bioinform.* **2018**, *19*, 524–536. [CrossRef]
20. Danezis, G.P.; Tsagkaris, A.S.; Camin, F.; Brusic, V.; Georgiou, C.A. Food Authentication: Techniques, Trends & Emerging Approaches. *TrAC—Trends Anal. Chem.* **2016**, *85*, 123–132. [CrossRef]
21. Aceto, M. The Use of ICP-MS in Food Traceability. In *Advances in Food Traceability Techniques and Technologies*; Elsevier: Amsterdam, The Netherlands, 2016; pp. 137–164, ISBN 978-0-08-100310-7.
22. Drivelos, S.A.; Georgiou, C.A. Multi-Element and Multi-Isotope-Ratio Analysis to Determine the Geographical Origin of Foods in the European Union. *TrAC—Trends Anal. Chem.* **2012**, *40*, 38–51. [CrossRef]
23. de Andrade, B.M.; Margalho, L.P.; Batista, D.B.; Lucena, I.O.; Kamimura, B.A.; Balthazar, C.F.; Brexó, R.P.; Pia, A.K.R.; Costa, R.A.S.; Cruz, A.G.; et al. Chemometric Classification of Brazilian Artisanal Cheeses from Different Regions According to Major and Trace Elements by ICP-OES. *J. Food Compos. Anal.* **2022**, *109*, 104519. [CrossRef]
24. Magdas, D.A.; Feher, I.; Cristea, G.; Voica, C.; Tabaran, A.; Mihaiu, M.; Cordea, D.V.; Bâlteanu, V.A.; Dan, S.D. Geographical Origin and Species Differentiation of Transylvanian Cheese. Comparative Study of Isotopic and Elemental Profiling vs. DNA Results. *Food Chem.* **2019**, *277*, 307–313. [CrossRef] [PubMed]
25. Tedesco, R.; Villoslada Hidalgo, M.D.C.; Vardè, M.; Kehrwald, N.M.; Barbante, C.; Cozzi, G. Trace and Rare Earth Elements Determination in Milk Whey from the Veneto Region, Italy. *Food Control* **2021**, *121*, 107595. [CrossRef]
26. Danezis, G.P.; Pappas, A.C.; Tsiplakou, E.; Pappa, E.C.; Zacharioudaki, M.; Tsagkaris, A.S.; Papachristidis, C.A.; Sotirakoglou, K.; Zervas, G.; Georgiou, C.A. Authentication of Greek Protected Designation of Origin Cheeses through Elemental Metabolomics. *Int. Dairy J.* **2020**, *104*, 104599. [CrossRef]
27. Nečemer, M.; Potočnik, D.; Ogrinc, N. Discrimination between Slovenian Cow, Goat and Sheep Milk and Cheese According to Geographical Origin Using a Combination of Elemental Content and Stable Isotope Data. *J. Food Compos. Anal.* **2016**, *52*, 16–23. [CrossRef]
28. Di Donato, F.; Foschi, M.; Vlad, N.; Biancolillo, A.; Rossi, L.; D'archivio, A.A. Multi-Elemental Composition Data Handled by Chemometrics for the Discrimination of High-Value Italian Pecorino Cheeses. *Molecules* **2021**, *26*, 6875. [CrossRef] [PubMed]
29. Santarcangelo, C.; Baldi, A.; Ciampaglia, R.; Dacrema, M.; Di Minno, A.; Pizzamiglio, V.; Tenore, G.C.; Daglia, M. Long-Aged Parmigiano Reggiano PDO: Trace Element Determination Targeted to Health. *Foods* **2022**, *11*, 172. [CrossRef] [PubMed]
30. Rodríguez-Bermúdez, R.; López-Alonso, M.; Miranda, M.; Fouz, R.; Orjales, I.; Herrero-Latorre, C. Chemometric Authentication of the Organic Status of Milk on the Basis of Trace Element Content. *Food Chem.* **2018**, *240*, 686–693. [CrossRef] [PubMed]
31. de Oliveira Filho, E.F.; Miranda, M.; Ferreiro, T.; Herrero-Latorre, C.; Castro Soares, P.; López-Alonso, M. Concentrations of Essential Trace and Toxic Elements Associated with Production and Manufacturing Processes in Galician Cheese. *Molecules* **2022**, *27*, 4938. [CrossRef]
32. Aceto, M.; Musso, D.; Calà, E.; Arieri, F.; Oddone, M. Role of Lanthanides in the Traceability of the Milk Production Chain. *J. Agric. Food Chem.* **2017**, *65*, 4200–4208. [CrossRef]
33. Danezis, G.P.; Georgiou, C.A. Elemental Metabolomics: Food Elemental Assessment Could Reveal Geographical Origin. *Curr. Opin. Food Sci.* **2022**, *44*, 100812. [CrossRef]

34. Urgeghe, P.P.; Piga, C.; Addis, M.; Di Salvo, R.; Piredda, G.; Scintu, M.F.; Wolf, I.V.; Sanna, G. SPME/GC-MS Characterization of the Volatile Fraction of an Italian PDO Sheep Cheese to Prevalent Lypolitic Ripening: The Case of Fiore Sardo. *Food Anal. Methods* **2012**, *5*, 723–730. [CrossRef]
35. Caredda, M.; Addis, M.; Ibba, I.; Leardi, R.; Scintu, M.F.; Piredda, G.; Sanna, G. Prediction of Fatty Acid Content in Sheep Milk by Mid-Infrared Spectrometry with a Selection of Wavelengths by Genetic Algorithms. *LWT* **2016**, *65*, 503–510. [CrossRef]
36. Idda, I.; Spano, N.; Ciulu, M.; Nurchi, V.M.; Panzanelli, A.; Pilo, M.I.; Sanna, G. Gas Chromatography Analysis of Major Free Mono- and Disaccharides in Milk: Method Assessment, Validation, and Application to Real Samples. *J. Sep. Sci.* **2016**, *39*, 4577–4584. [CrossRef] [PubMed]
37. Pulinas, L.; Spanu, C.; Idda, I.; Ibba, I.; Nieddu, G.; Virdis, S.; Scarano, C.; Piras, F.; Spano, N.; Sanna, G.; et al. Production of Farmstead Lactose-Free Pecorino Di Osilo and Ricotta Cheeses from Sheep's Milk. *Ital. J. Food Saf.* **2017**, *6*, 1. [CrossRef]
38. Idda, I.; Spano, N.; Addis, M.; Galistu, G.; Ibba, I.; Nurchi, V.M.; Pilo, M.I.; Scintu, M.F.; Piredda, G.; Sanna, G. Optimization of a Newly Established Gas-Chromatographic Method for Determining Lactose and Galactose Traces: Application to Pecorino Romano Cheese. *J. Food Compos. Anal.* **2018**, *74*, 89–94. [CrossRef]
39. Piga, C.; Urgeghe, P.P.; Piredda, G.; Scintu, M.F.; Sanna, G. Assessment and Validation of Methods for the Determination of γ-Glutamyltransferase Activity in Sheep Milk. *Food Chem.* **2009**, *115*, 1519–1523. [CrossRef]
40. Piga, C.; Urgeghe, P.P.; Piredda, G.; Scintu, M.F.; Sanna, G. Analytical Methods for the Evaluation of α-l-Fucosidase Activity in Sheep Milk. *Food Anal. Methods* **2010**, *3*, 17–21. [CrossRef]
41. Piga, C.; Urgeghe, P.P.; Piredda, G.; Scintu, M.F.; Di Salvo, R.; Sanna, G. Thermal Inactivation and Variability of γ-Glutamyltransferase and α-l-Fucosidase Enzymatic Activity in Sheep Milk. *LWT—Food Sci. Technol.* **2013**, *54*, 152–156. [CrossRef]
42. Dedola, A.S.; Piras, L.; Addis, M.; Pirisi, A.; Piredda, G.; Mara, A.; Sanna, G. New Analytical Tools for Unmasking Frauds in Raw Milk-Based Dairy Products: Assessment, Validation and Application to Fiore Sardo PDO Cheese of a RP-HPLC Method for the Evaluation of the α-l-Fucosidase Activity. *Separations* **2020**, *7*, 40. [CrossRef]
43. Caredda, M.; Addis, M.; Ibba, I.; Leardi, R.; Scintu, M.F.; Piredda, G.; Sanna, G. Building of Prediction Models by Using Mid-Infrared Spectroscopy and Fatty Acid Profile to Discriminate the Geographical Origin of Sheep Milk. *LWT* **2017**, *75*, 131–136. [CrossRef]
44. Zazzu, C.; Addis, M.; Caredda, M.; Scintu, M.F.; Piredda, G.; Sanna, G. Biogenic Amines in Traditional Fiore Sardo Pdo Sheep Cheese: Assessment, Validation and Application of an Rp-Hplc-Dad-Uv Method. *Separations* **2019**, *6*, 11. [CrossRef]
45. Guiso, M.F.; Battacone, G.; Canu, L.; Deroma, M.; Langasco, I.; Sanna, G.; Tsiplakou, E.; Pulina, G.; Nudda, A. Essential and Toxic Mineral Content and Fatty Acid Profile of Colostrum in Dairy Sheep. *Animals* **2022**, *12*, 2730. [CrossRef] [PubMed]
46. Spano, N.; Bortolu, S.; Addis, M.; Langasco, I.; Mara, A.; Pilo, M.I.; Sanna, G.; Urgeghe, P.P. An Analytical Protocol for the Differentiation and the Potentiometric Determination of Fluorine-Containing Fractions in Bovine Milk. *Molecules* **2023**, *28*, 1349. [CrossRef] [PubMed]
47. Kowalczyk, E.; Givelet, L.; Amlund, H.; Sloth, J.J.; Hansen, M. Risk Assessment of Rare Earth Elements, Antimony, Barium, Boron, Lithium, Tellurium, Thallium and Vanadium in Teas. *EFSA J.* **2022**, *20*, e200410. [CrossRef] [PubMed]
48. Manuelian, C.L.; Currò, S.; Penasa, M.; Cassandro, M.; De Marchi, M. Characterization of Major and Trace Minerals, Fatty Acid Composition, and Cholesterol Content of Protected Designation of Origin Cheeses. *J. Dairy Sci.* **2017**, *100*, 3384–3395. [CrossRef]
49. Coni, E.; Bocca, B.; Caroli, S. Minor and Trace Element Content of Two Typical Italian Sheep Dairy Products. *J. Dairy Res.* **1999**, *66*, 589–598. [CrossRef]
50. Danezis, G.; Theodorou, C.; Massouras, T.; Zoidis, E.; Hadjigeorgiou, I.; Georgiou, C.A. Greek Graviera Cheese Assessment through Elemental Metabolomics—Implications for Authentication, Safety and Nutrition. *Molecules* **2019**, *24*, 670. [CrossRef]
51. Recio, I.; de la Fuente, M.A.; Juárez, M.; Ramos, M. Bioactive Components in Sheep Milk. In *Bioactive Components in Milk and Dairy Products*; John Wiley & Sons, Ltd.: Hoboken, NJ, USA, 2009; pp. 83–104, ISBN 978-0-8138-2150-4.
52. de la Fuente, M.A.; Olano, A.; Juârez, M. Distribution of Calcium, Magnesium, Phosphorus, Zinc, Manganese, Copper and Iron between the Soluble and Colloïdal Phases of Ewe's and Goat's Milk. *Lait* **1997**, *77*, 515–520. [CrossRef]
53. Polychroniadou, A.; Vafopoulou, A. Variations of Major Mineral Constituents of Ewe Milk during Lactation. *J. Dairy Sci.* **1985**, *68*, 147–150. [CrossRef]
54. Pellegrini, O.; Remeuf, F.; Rivemal, M. Evolution of Physico-Chemical Characteristics and Renneting Properties of Ewe's Milk Collected in the "Roquefort Area". *Lait* **1994**, *74*, 425–442. [CrossRef]
55. Martín-Hernández, M.C.; Amigo, L.; Martín-Álvarez, P.J.; Juárez, M. Differentiation of Milks and Cheeses According to Species Based on the Mineral Content. *Eur. Food Res. Technol.* **1992**, *194*, 541–544. [CrossRef]
56. Timlin, M.; Tobin, J.T.; Brodkorb, A.; Murphy, E.G.; Dillon, P.; Hennessy, D.; O'donovan, M.; Pierce, K.M.; O'callaghan, T.F. The Impact of Seasonality in Pasture-Based Production Systems on Milk Composition and Functionality. *Foods* **2021**, *10*, 607. [CrossRef]
57. Holt, C. Milk Salts | Interaction with Caseins. In *Encyclopedia of Dairy Sciences*, 2nd ed.; Fuquay, J.W., Ed.; Academic Press: Cambridge, MA, USA, 2011; pp. 917–924, ISBN 978-0-12-374407-4.
58. Fox, P.; McSweeney, P. *Advanced Dairy Chemistry*; Springer: Berlin/Heidelberg, Germany, 2009; pp. 351–389, ISBN 978-0-387-84864-8.

59. Woodrow, J.; Sharpe, C.; Fudge, N.; Hoff, A.; Gagel, R.; Kovacs, C.S. Calcitonin Plays a Critical Role in Regulating Skeletal Mineral Metabolism during Lactation. *Endocrinology* **2006**, *147*, 4010–4021. [CrossRef] [PubMed]
60. Liesegang, A.; Eicher, R.; Sassi, M.-L.; Risteli, J.; Kraenzlin, M.; Riond, J.-L.; Wanner, M. Biochemical Markers of Bone Formation and Resorption Around Parturition and during Lactation in Dairy Cows with High and Low Standard Milk Yields. *J. Dairy Sci.* **2000**, *83*, 1773–1781. [CrossRef]
61. Gulati, A.; Galvin, N.; Lewis, E.; Hennessy, D.; O'Donovan, M.; McManus, J.J.; Fenelon, M.A.; Guinee, T.P. Outdoor Grazing of Dairy Cows on Pasture versus Indoor Feeding on Total Mixed Ration: Effects on Gross Composition and Mineral Content of Milk during Lactation. *J. Dairy Sci.* **2018**, *101*, 2710–2723. [CrossRef] [PubMed]
62. Hettinga, K. Lactose in the Dairy Production Chain. In *Lactose: Evolutionary Role, Health Effects, and Applications*; Elsevier: Amsterdam, The Netherlands; Academic Press: Cambridge, MA, USA, 2019; pp. 231–266, ISBN 978-0-12-811720-0.
63. Frédéric, G. The Minerals of Milk. *Reprod. Nutr. Dev.* **2005**, *45*, 473–483. [CrossRef]
64. Ebringer, L.; Ferencik, M.; Krajcovic, J. Beneficial Health Effects of Milk and Fermented Dairy Products. *Folia Microbiol.* **2008**, *53*, 378–394. [CrossRef] [PubMed]
65. Theobald, H. Dietary Calcium and Health. *Nutr. Bull.* **2005**, *30*, 237–277. [CrossRef]
66. U.S. Department of Health & Human Services. National Institutes of Health Nutrient Recommendations and Databases. 2019. Available online: https://ods.od.nih.gov/HealthInformation/nutrientrecommendations.aspx (accessed on 15 January 2024).
67. European Union Council Directive 90/496/EEC on Nutrition Labelling Rules of Foodstuffs. 1990. Available online: https://eur-lex.europa.eu/legal-content/EN/ALL/?uri=celex:31990L0496 (accessed on 15 January 2024).
68. European Union Regulation (EC) No. 1924/2006 on Nutrition and Health Claims Made on Foods. 2006. Available online: https://eur-lex.europa.eu/LexUriServ/LexUriServ.do?uri=OJ:L:2006:404:0009:0025:En:PDF (accessed on 15 January 2024).
69. European Union Regulation (EC) No. 432/2012 on the Compilation of a List of Permitted Health Claims on Food, Other Than Those Referring to the Reduction of Disease Risks and to the Development and Health of Children. 2012. Available online: https://eur-lex.europa.eu/LexUriServ/LexUriServ.do?uri=OJ:L:2012:136:0001:0040:en:PDF (accessed on 15 January 2024).
70. Consorzio per la Tutela del Formaggio Pecorino Romano, Disciplinare Pecorino Romano DOP. 2009. Available online: https://www.pecorinoromano.com/pecorino-romano/disciplinare-e-norme (accessed on 15 January 2024).
71. Consorzio per la Tutela del Formaggio Pecorino Sardo, Disciplinare Pecorino Sardo DOP. 1996. Available online: https://www.politicheagricole.it/flex/files/9/b/b/D.2baf7e91401d2f673d3e/Disciplinare_pecorino_sardo.pdf (accessed on 15 January 2024).
72. Subramanian, A.; Rodriguez-Saona, L. Chemical and instrumental approaches to cheese analysis. *Adv. Food Nutr. Res.* **2010**, *59*, 167–213. [CrossRef]
73. Mara, A.; Deidda, S.; Caredda, M.; Ciulu, M.; Deroma, M.; Farinini, E.; Floris, I.; Langasco, I.; Leardi, R.; Pilo, M.I.; et al. Multi-Elemental Analysis as a Tool to Ascertain the Safety and the Origin of Beehive Products: Development, Validation, and Application of an ICP-MS Method on Four Unifloral Honeys Produced in Sardinia, Italy. *Molecules* **2022**, *27*, 2009. [CrossRef]
74. Astolfi, M.L.; Conti, M.E.; Marconi, E.; Massimi, L.; Canepari, S. Effectiveness of Different Sample Treatments for the Elemental Characterization of Bees and Beehive Products. *Molecules* **2020**, *25*, 4263. [CrossRef]
75. Currie, L.A. Nomenclature in Evaluation of Analytical Methods Including Detection and Quantification Capabilities (IUPAC Recommendations 1995). *Anal. Chim. Acta* **1999**, *391*, 105–126. [CrossRef]
76. Appendix F, Guidelines for Standard Method Performance Requirements, AOAC Official Methods of Analysis. 2016. Available online: https://www.aoac.org/wp-content/uploads/2019/08/app_f.pdf (accessed on 15 January 2024).
77. Leardi, R.; Melzi, C.; Polotti, G. CAT (Chemometric Agile Tool Software). Available online: http://gruppochemiometria.it/index.php/software (accessed on 15 January 2024).

molecules

MDPI

Article

Simultaneous Determination of Seven Pesticides and Metabolite Residues in Litchi and Longan through High-Performance Liquid Chromatography-Tandem Mass Spectrometry with Modified QuEChERS

Siwei Wang [1,2], Xiaonan Wang [1], Qiang He [3], Haidan Lin [3], Hong Chang [1], Haibin Sun [1] and Yanping Liu [1,*]

[1] Institute of Plant Protection, Guangdong Academy of Agricultural Sciences, Key Laboratory of Green Prevention and Control on Fruits and Vegetables in South China Ministry of Agriculture and Rural Affairs, Guangdong Provincial Key Laboratory of High Technology for Plant Protection, Guangzhou 510640, China

[2] Guangdong Engineering Research Center for Insect Behavior Regulation, South China Agricultural University, Guangzhou 510642, China

[3] Guangdong Quality Safety Center of Agricultural Products, Department of Agriculture and Rural Affairs of Guangzhou, Guangzhou 510230, China

* Correspondence: liuliuyp@tom.com; Tel.: +86-020-87597555

Abstract: This study established a QuEChERS high-performance liquid chromatography/tandem triple-quadrupole mass spectrometry method for determining azoxystrobin, pyraclostrobin, picoxystrobin, difenoconazole, chlorantraniliprole, imidacloprid, and cyantraniliprole and its metabolite (IN-J9Z38) in litchi and longan, and applied this method to the real samples. The residues in samples were extracted with acetonitrile and purified with nano-ZrO_2, C_{18}, and PSA. The samples were then detected with multireactive ion monitoring and electrospray ionization in the positive ion mode and quantified using the external matrix-matched standard method. The results showed good linearities for the eight analytes in the range of 1–100 μg/L, with correlation coefficients (r^2) of >0.99. The limit of quantification was 1–10 μg/kg, and the limit of detection was 0.3–3 μg/kg. Average recovery from litchi and longan was 81–99%, with the relative standard deviation of 3.5–8.4% at fortified concentrations of 1, 10, and 100 μg/kg. The developed method is simple, rapid, efficient, and sensitive. It allowed the rapid screening, monitoring, and confirming of the aforementioned seven pesticides and a metabolite in litchi and longan.

Keywords: QuEChERS; HPLC-MS/MS; litchi and longan; pesticide and metabolite; residue

Citation: Wang, S.; Wang, X.; He, Q.; Lin, H.; Chang, H.; Sun, H.; Liu, Y. Simultaneous Determination of Seven Pesticides and Metabolite Residues in Litchi and Longan through High-Performance Liquid Chromatography-Tandem Mass Spectrometry with Modified QuEChERS. *Molecules* **2022**, *27*, 5737. https://doi.org/10.3390/molecules27175737

Academic Editor: Stefano Dall'Acqua

Received: 11 August 2022
Accepted: 1 September 2022
Published: 5 September 2022

1. Introduction

Litchi and longan are both vital economic fruit trees in tropical and subtropical regions of China. Their annual output in China ranks highest in the world [1–4]. Litchi is known as the "king of fruits", and longan is known as "southern ginseng" in China [5–8]. The high temperature and humidity in the planting areas of China make litchi and longan trees vulnerable to diseases and pests. At present, 140 pesticide products in litchi have been registered in China, including 48 active ingredients (10 insecticides, 23 fungicides, 1 herbicide, and 14 plant growth regulators) [9]. Seven pesticide products have been registered in longan, of which only 6 are active ingredients (2 insecticides, 1 fungicide, and 3 plant growth regulators) [9]. Most registered pesticides are pyrethroids, organophosphorus, triazole, and carbamate, and these chemical pesticides have been used for many years in litchi and longan orchards. The annual high-dose and frequent usage of pesticides has led to the prominent problems of resistance, pesticide residue pollution, and dose exceeding the standard in litchi and longan orchards [10]. The repeated use of chemical pesticides causes not only toxicity to fruits, but also environmental pollution and harm to non-target biological entities. For example, the rate of acute oral and contact toxicity of pyrethroid and organophosphorus insecticides to bees, fish, and birds is high [11–13]. An increasing number of studies have proven that fungicides could also harm the health of non-target organisms such as bees [14–16]. Triazole fungicides are classified as "potential human carcinogens" by the US Environmental Protection Agency [17,18]. Consumers are also increasingly concerned about the health and pollution problems caused by pesticides and their metabolite residues. The toxicity of some pesticide metabolites may even be higher than that of the parent pesticide [19]. The fruit quality and safety and environmental problems caused by pesticides deserve attention. Hence, establishing a method for the simultaneous determination of pesticides and their metabolites in litchi and longan is valuable.

At present, the main detection and analytical methods for the aforementioned pesticides are gas chromatography-electron capture detection (GC-ECD), gas chromatography-mass spectrometry (GC-MS), high-performance liquid chromatography (HPLC), and HPLC-mass spectrometry (HPLC-MS/MS) [20–32]. The main matrix includes vegetables such as broccoli and zucchini, fruits such as pears and apples, and traditional Chinese medicine [20–32]. Few studies have reported on multiresidue analysis of pesticides and their metabolites in litchi and longan. Common sample purification methods include solid phase extraction (SPE), liquid–liquid distribution extraction, and QuEChERS [20–32]. SPE is a tedious and time-consuming method involving the use of a large amount of organic solvents and nitrogen blowing, which is relatively cumbersome. Many types of commercial SPE columns are available, but selecting suitable SPE column for multiresidue analysis is difficult. The dispersive liquid–liquid extraction method requires a large amount of organic solvents and is prone to emulsification, which affects the recovery of target analytes. QuEChERS can achieve better purification with the addition of a small amount of adsorbent to the extraction solution to adsorb impurities. The operation of QuEChERS, which is currently a common purification method for fruits and vegetables, is fast and simple [33–36].

The main objectives of this paper are to establish and validate a fast and sensitive QuEChERS method employing a combination of nano-ZrO_2, C_{18}, and PSA for sample pretreatment before HPLC-MS/MS analysis for the simultaneous quantification of 7 pesticides and 1 metabolite in litchi and longan.

2. Materials and Methods

2.1. Materials and Reagents

The standards of azoxystrobin (99.5% purity), difenoconazole (99.0% purity), chlorantraniliprole (97.8% purity), pyraclostrobin (99.5% purity), and imidacloprid (98.5% purity) were purchased from Dr Ehrenstorfer GmbH (Wesel, Germany). The standard of picoxystrobin (98.4% purity) was purchased from Chem Service (West Chester, PA, USA). The standards of cyantraniliprole (98.0% purity) and its metabolite IN-J9Z38 (97.8% purity) were purchased from Toronto Research Chemicals (Toronto, Canada) and Hangzhou Trylead Chemical Technology Co., Ltd. (Hangzhou, China). HPLC grade acetonitrile (MeCN) and methanol were obtained from Fisher (Pittsburgh, PA, USA). Chromatographically pure formic acid was obtained from Fluka (Seeize, Germany). Analytical grade anhydrous magnesium sulfate (anhydrous $MgSO_4$) and sodium chloride (NaCl) were obtained from Sinopharm Chemical Reagents Co., Ltd. (Shanghai, China). The octadecylsilane (C_{18}, 40 µm) adsorbent was purchased from Agela Technologies Inc. (Tianjin, China). The graphitized carbon black (GCB, 120–400 MESH) and primary secondary amine (PSA, 40–60 µm) were obtained from ANPEL Laboratory Technologies Inc. (Shanghai, China). Nano-ZrO_2 (99.99%, ≤100 nm) was from Aladdin Biochemical Technology Co., Ltd. (Shanghai, China), and multi-walled carbon nanotubes (MWCNTs; 95%, 10–30 µm, 8 nm) were from Nanjing XFNANO Materials Technology Co., Ltd. Moreover, the ultra-pure water was prepared using a Milli-Q Integral Water Purification System (Millipore Corporation; Burlington, MA, USA). The filter membrane (13 mm × 0.22 µm) was obtained from Ameritech Science and Technology Ltd. (Chicago, IL, USA). The high-speed refrigeration centrifuge model (GTR22-1) was also used (Beili Medicine Centrifuge Factory; Beijing, China).

2.2. Sample Pretreatment

The mashed litchi or longan sample (10.0 g) was weighed and extracted using 10 mL MeCN. The mixture was shaken for 2 min. Subsequently, 2 g NaCl and 4 g anhydrous $MgSO_4$ were added. After mixing for an additional 1 min, the mixture was placed on a horizontal oscillator for 10 min. Samples were centrifuged for 5 min at 5204× *g*. Next, 2 mL of the supernatant was transferred into a 5 mL centrifuge tube containing 300 mg anhydrous $MgSO_4$, 25 mg C_{18}, 25 mg PSA, and 10 mg nano-ZrO_2 for cleanup. After shaking for 1 min, samples were again centrifuged for 5 min at 5204× *g*, with supernatants then being passed through a 0.22 μm nylon syringe filter and transferred to an autosampler vial to conduct HPLC–MS/MS analyses.

2.3. Instrumental Parameters

A Shimadzu LC-20A HPLC system was used for separating the target analytes on a InfinityLab Poroshell 120 SB-C18 column (Dim:75 mm × 2.1 mm, 2.7 μm particle size, Agilent, Palo Alto, USA) by using a column oven at 35 °C. The separation was performed through gradient elution with A (0.1% of formic acid aqueous solution) and B (MeCN) as the mobile phase, and the flow rate was kept constant (0.3 mL/min) during the complete analysis process. The gradient program was: 0–2 min 80% A–5% A, 2–3.5 min 5% A, 3.5–4.5 min 20% A, 4.5–6 min 20% A. Injection (5 μL) was conducted using an autosampler.

A triple quadrupole mass spectrometer (Shimadzu 8045; Shimadzu; Kyoto, Japan) equipped with electrospray ionization in the positive ion mode (ESI^+) was used to quantify the target analytes. The oven temperature was set at 350 °C, the desolvation line was set at 250 °C, the temperature of the heating block was 400 °C, nitrogen was used as a nebulizer and collision gas, and multireactive ion monitoring (MRM) was selected to analyze the target analytes with a dwell time of 80 ms. The optimal precursor ions, product ions, collision energies, and other instrument parameters for each analyte were acquired by directly infusing each target pesticide at a concentration of 100 μg/L standard solution into the ion source in the instrument. All other relevant MS parameters are listed in Table 1.

Table 1. Mass parameters of eight targeted compounds.

Compound	Precursor Ion (*m*/*z*)	Daughter Ion (*m*/*z*)	CE/eV	Q1Pre (V)	Q3Pre (V)	Retention Time RT/min
Azoxystrobin	404.20	344.10 *, 329.10	−25, −30	−15, −19	−13, −22	3.25
Pyraclostrobin	388.15	163.10 *, 133.05	−24, −36	−29, −14	−17, −24	3.62
Picoxystrobin	368.00	145.10 *, 117.25	−23, −38	−28, −29	−27, −22	3.51
Difenoconazole	406.10	251.00 *, 337.00	−26, −18	−12, −19	−30, −24	3.59
Chlorantraniliprole	484.00	285.90 *, 453.35	−12, −14	−23, −18	−29, −23	2.97
Cyantraniliprole	475.31	285.95 *, 444.10	−10, −19	−30, −24	−15, −18	3.33
IN-J9Z38	457.10	299.00 *, 188.00	−27, −35	−20, −17	−26, −14	3.42
Imidacloprid	256.10	209.05 *, 175.10	−14, −18	−18, −19	−22, −18	2.40

* Quantitative ion.

2.4. Method Validation

Recovery rate, linearity, and limit of quantification (LOQ) values were used to assess the accuracy and reliability of the developed techniques. The standards produced a linear result between 1 and 100 μg/L. Fortified blank food samples at concentrations of 1, 10, and 100 μg/kg were used to evaluate the recovery rate, and each spiked level was replicated six times, whereas the corresponding relative standard deviations (RSDs) represented the method precision. Based on the guidelines of SANTE/11813/2017, the LOQ for pesticide was considered as the lowest spiked level in the matrix.

3. Results and Discussion

3.1. Optimization of Chromatographic Separating Column

The separation effects of the two liquid chromatographic columns on the target analytes were compared. A: Shimadzu Shim-pack GIST-HP C18 (Dim: 50 mm × 2.1 mm, 3.0 μm particle size), and B: InfinityLab Poroshell C18 (Dim: 75 mm × 2.1 mm, 2.7 μm particle size). Chromatographic column A was a short column which has the advantage of a short analysis time, but the peak symmetry of the 8 target analytes was poor. Chromatographic column B exhibits good response values and peak symmetry. Hence, InfinityLab Poroshell C18 was considered to be the most desirable chromatographic column. The relevant chromatographs are shown in Figure 1.

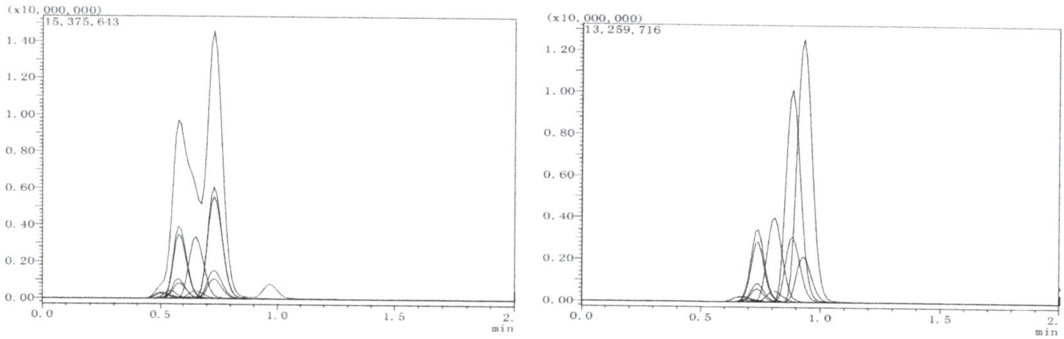

(A) (B)

Figure 1. The total ion chromatogram of eight targeted compounds separated by the two different columns (lateral axis is time, and longitudinal axis is response intensity). (**A**) Shim-pack GIST-HP C18, 50 mm × 2.1 mm, 3.0 μm; (**B**) InfinityLab Poroshell C18, 75 mm × 2.1 mm, 2.7 μm.

3.2. Optimization of Mobile Phase

The mobile phase can influence the resolution effect, response value, and retention time of the target analytes. A shorter retention time is desirable if baseline separation can be achieved. Using the 75-mm InfinityLab Poroshell C18 chromatographic column, the effects of mobile phases A (methanol and 0.1% formic acid aqueous solution), B (MeCN and 0.1% formic acid aqueous solution), C (methanol and water), and D (MeCN and water) on the resolution and response intensity of the 8 target analytes were investigated. The results showed that the mobile phases C and D without formic acid had a great effect on the resolution, which led to asymmetric peak and peak broadened. The addition of ammonium formate, and formic acid to the aqueous phase is a common means to improve the chromatographic peak shape, the instrument response value, and ionization efficiency. Generally, the use of an acidic mobile phase is conducive to mass spectrometry detection in positive ion mode and formic acid is one of the most commonly used modifiers. With the addition of 0.1% formic acid to the aqueous phase, the mobile phase components of A and B promoted the formation of $[M + H]^+$ ion peaks and improved the analytical sensitivity of the target analyte. The response of B to the target analyte was relatively higher. Hence, B (MeCN and 0.1% formic acid aqueous solution) was chosen as the mobile phase. The relevant chromatographs are shown in Figure 2.

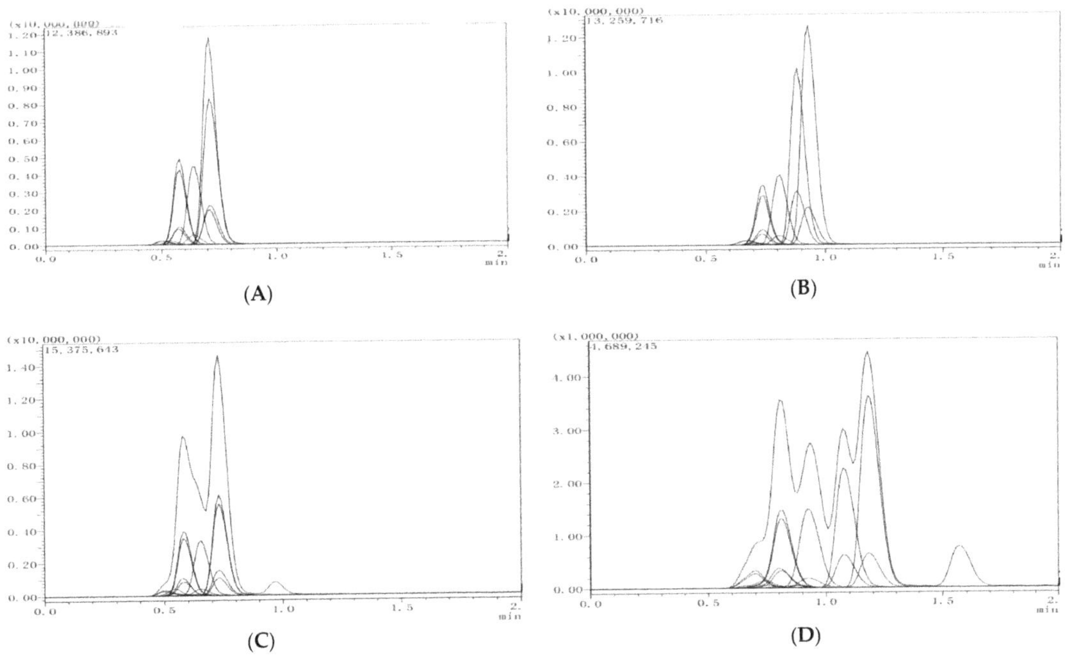

Figure 2. The chromatogram of eight targeted compounds separated by the different mobile phase components (lateral axis is time, and longitudinal axis is response intensity). (**A**) Methanol-water containing 0.1% formic acid; (**B**) acetonitrile-water containing 0.1% formic acid; (**C**) methanol-water; (**D**) acetonitrile-water.

3.3. Comparison of Constant and Gradient Elution

MeCN and 0.1% formic acid aqueous solution were selected as the mobile phase. The effects of constant elution (A) and gradient elution (B) on the resolution and response intensity of the target compounds were investigated. The peaks of the 8 target analytes were symmetrical, and no peak broadening phenomenon was observed at the gradient elution mode. The resolution of each target compound and the accuracy of quantitative results could be good. The relevant chromatographs are shown in Figure 3.

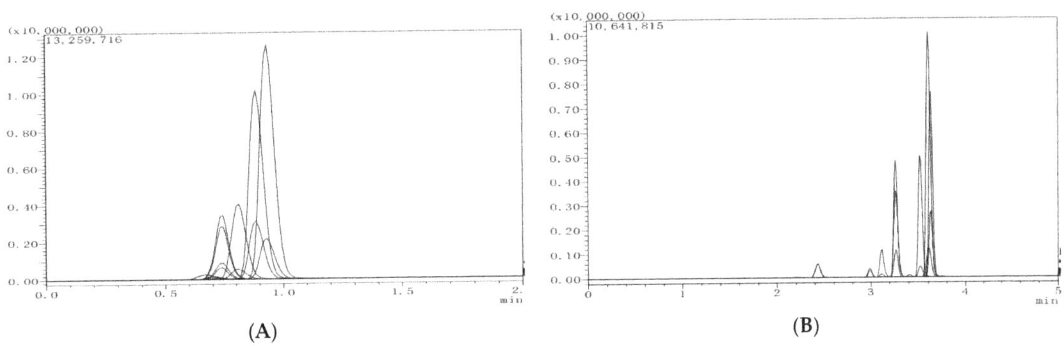

Figure 3. The chromatogram of eight targeted compounds separated by the constant elution (**A**) and gradient elution, (**B**) of mobile phase (lateral axis is time, and longitudinal axis is response intensity).

3.4. Optimization of MS/MS

The eight target analytes contained strong negative dielectrics such as oxygen and nitrogen atoms, which could form adduct ions ($[M + H]^+$) with the hydrogen ions in the spray droplets under the positive ion mode (ESI^+). The stability $[M + H]^+$ of each pesticide was obtained using MS positive ion detection of 100 μg/kg of standard solution. To improve the ionization efficiency of the target analyte, the relevant parameters in the source were optimized. In the positive ion mode, the scanning range was set according to the relative molecular weight of the target compound, Q3 full scan on the compound was carried out, and stable $[M + H]^+$ molecular ions were obtained through primary mass spectrometry scanning, and the parent ions were determined. Then, the target compound was scanned to obtain product ions. The CE, Q1 pre, Q3 pre, and product ions of the 8 target compounds were optimized in the MRM mode. Finally, the mass spectrum conditions of the target compounds were determined.

3.5. Optimization of Extraction Solvents

According to the relevant literature, the most commonly used extraction solvents for the QuEChERS method are ethyl acetate, acetone, and acetonitrile. Acetone cannot be separated from water in the absence of a non-polar solvent, but it can easily extract the pigment and other impurities from the matrix. In particular, the litchi pericarp contains more pigments, flavonoids, polyphenols, and other impurities, and the co-extraction phenomenon is more severe. The extraction efficiency of ethyl acetate is low when extracting polar pesticides, and ethyl acetate can be easily emulsified. Acetonitrile can be easily separated from water by addition of NaCl, and it extracts fewer impurities. Acetonitrile is the most commonly used extraction solvent in the QuEChERS method [28]. Hence, acetonitrile was chosen as the extraction solvent.

3.6. Optimization of Purification Adsorbents

Sample purification is a very important step that could avoid any potential contamination of the chromatographic column and MS detection. The QuEChERS method has the advantage of less organic solvent consumption, good reproducibility, and high sensitivity, that is widely applied for multi-residue analysis in fruits and vegetables. Hence, selecting appropriate adsorbents is of great importance for quantitative accuracy.

The adsorption effect and recovery rate of 8 target compounds with the five adsorbents (MWCNTs, nano-ZrO_2, PSA, C_{18}, and GCB) were investigated. The MWCNTs belongs a new conductive carbon nanometer material with a larger specific surface area, and with a good purification effect on pigments. In this study, the recoveries of the 8 target compounds were low (ranged from 1% to 56% except picoxystrobin (88%) and imidacloprid (77%)) when the dosage of MWCNTs was 5–10 mg, indicating that the MWCNTs adsorbent has a strong adsorption capacity for the target compound and is not easy to desorb. Although GCB could significantly remove the pigment from the sample, the recovery rate was low due to the strong adsorption of GCB on difenoconazole, chlorantraniliprole, cyantraniliprole, and IN-J9Z38. The nano-ZrO_2, PSA, and C18 could satisfy the requirement of a recovery range of 70−120%, exhibited good recoveries of the target compounds (90–103%, Figure 4).

3.7. Optimization of Adsorbent Dosage

The purification effect of a single adsorbent for co-extracted impurities is considerably weaker than that of combined adsorbents, especially for litchi and longan, which are relatively complex matrices. Therefore, the combination of different adsorbents was tested to establish the most effective approach to producing a purified sample with satisfactory recovery. The advantages of C_{18} adsorbent are its large specific surface area, which can effectively remove non-polar interferences such as fats and lipids. PSA is a weak anion exchange filler that can effectively remove fatty acids, pigments, sugars, and other substances from the matrix. Nano-ZrO_2 has small particle size, large specific surface area, and removes lipophilic impurities. The recovery of the target compounds ranged from 86% to 103%

when the dosage of PSA, C_{18}, and nano-ZrO_2 was in the range of 25–50 mg, 25–50 mg, and 10–20 mg (Table 2). As such, a combination of 25 mg C_{18} + 25 mg PSA + 10 mg nano-ZrO_2 was selected as the adsorbent of choice for these analyses.

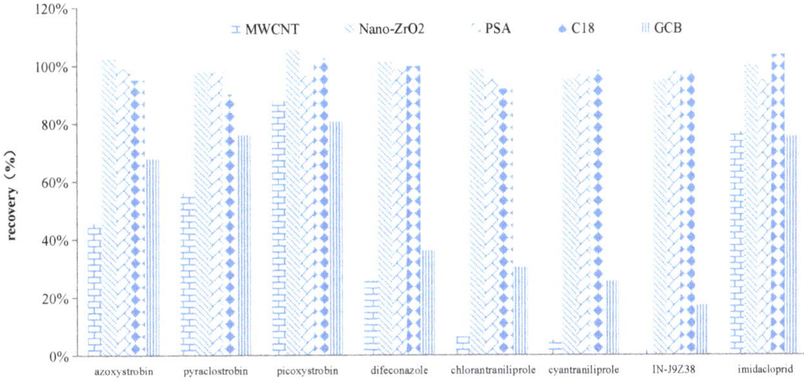

Figure 4. Absorption efficiency of five different adsorbent for eight targeted compounds.

Table 2. Effect of adsorbent dosage on the recovery of eight targeted compounds in litchi and longan samples (n = 5).

Different Adsorbent Combination	Recovery/%							
	Azoxystrobin	Pyraclostrobin	Picoxystrobin	Difenoconazole	Chlorantraniliprole	Cyantraniliprole	IN-J9Z38	Imidacloprid
25 mg PSA + 25 mg C_{18}	100 ± 2.5	100 ± 2.0	100 ± 3.5	98 ± 4.0	100 ± 2.0	99 ± 2.7	98 ± 5.1	100 ± 2.1
25 mg PSA + 10 mg nano-ZrO_2	96 ± 3.2	96 ± 4.0	99 ± 2.5	102 ± 1.5	97 ± 3.5	98 ± 4.0	97 ± 4.0	99 ± 3.8
25 mg C_{18} + 10 mg nano-ZrO_2	99 ± 3.8	102 ± 2.0	100 ± 2.1	99 ± 3.5	96 ± 4.9	97 ± 3.2	91 ± 1.5	100 ± 2.7
50 mg PSA + 50 mg C_{18}	97 ± 4.2	99 ± 1.0	102 ± 2.7	103 ± 3.1	102 ± 1.5	100 ± 1.5	100 ± 2.5	103 ± 2.5
50 mg PSA + 20 mg nano-ZrO_2	88 ± 2.0	97 ± 2.5	93 ± 2.5	96 ± 3.5	94 ± 3.1	95 ± 3.1	90 ± 1.5	97 ± 3.6
50 mg C_{18} + 20 mg nano-ZrO_2	89 ± 1.0	95 ± 5.0	92 ± 4.4	93 ± 3.8	94 ± 3.0	91 ± 6.8	90 ± 4.0	99 ± 3.0
25 mg PSA + 25 mg C_{18} + 10 mg nano-ZrO_2	90 ± 2.1	91 ± 2.5	90 ± 3.5	90 ± 3.5	92 ± 2.5	90 ± 1.5	88 ± 2.0	94 ± 5.0
50 mg PSA + 50 mg C_{18} + 20 mg nano-ZrO_2	87 ± 7.6	94 ± 3.6	88 ± 3.6	89 ± 2.3	90 ± 1.5	88 ± 5.1	86 ± 3.1	93 ± 4.5

3.8. Matrix Effect

The ME refers to the influence of components other than analytes in the sample on the response value of the analytes [33]. ME is an important factor affecting the accuracy of HPLC-MS/MS quantitative results [37]. ME (%) = $[(m_{matrix}/m_{solvent}) - 1)] \times 100\%$, where m_{matrix} is the slope of the matrix matching standard curve, and $m_{solvent}$ is the slope of the pure solvent standard curve. A 100 mg/L mixed standard solution was diluted with litchi and longan matrix purification solutions stepwise to prepare the matrix standard curve. A positive ME indicates the matrix enhancement effect, and the matrix can improve the response of the target. A negative ME indicates the matrix inhibition effect, and the matrix can reduce the response of the target [34]. The ME is divided into 3 grades according to the absolute value of ME. When the absolute ME value is 0–20%, the ME is weak; when the absolute ME value is 20–50%, the ME is medium; and when the absolute ME value is >50%, the ME is strong [38–41]. Table 3 shows that only the ME of pyraclostrobin was in the range of 0–20% in longan, indicating the presence of a weak ME. Azoxystrobin, picoxystrobin, difenoconazole, chlorantraniliprole, cyantraniliprole, and IN-J9Z38 in longan have a medium ME (0.5–50%). In this study, the matrix matching standard solution was used to correct the ME.

Table 3. Linear equations and matrix effect of eight targeted compounds in litchi and longan samples (n = 5).

Compound	Sample	Linear Range /μg/L	Linear Equation	Correlation Coefficient /R^2	ME [a] (%)
Azoxystrobin	solvent	1–100	$Y = 1.3988 \times 10^8X + 114{,}089$	0.9988	
	Litchi	1–100	$Y = 5.73765 \times 10^7X - 11{,}998.3$	0.9989	−59.03
	Longan	1–100	$Y = 7.0 \times 10^7X + 900{,}000$	0.998	−50.01
Pyraclostrobin	solvent	1–100	$Y = 2.21447 \times 10^8C + 340{,}737$	0.9906	
	Litchi	1–100	$Y = 5.56824 \times 10^7X + 853{,}513$	0.9996	−74.92
	Longan	1–100	$Y = 2.22603 \times 10^8X - 499{,}085$	0.9902	0.52
Picoxystrobin	solvent	1–100	$Y = 1.40371 \times 10^8X - 129{,}785$	0.9998	
	Litchi	1–100	$Y = 2.59173 \times 10^7X - 7809.87$	0.9998	−81.52
	Longan	1–100	$Y = 7.20201 \times 10^7X - 97{,}424.8$	0.997	−48.67
Difenoconazole	solvent	1–100	$Y = 3.14863 \times 10^8X - 664{,}206$	0.9978	
	Litchi	1–100	$Y = 1.39059 \times 10^8X + 132{,}133$	0.9992	−55.82
	longan	1–100	$Y = 1.0 \times 10^8X + 2 \times 10^7$	0.9901	−48.19
Chlorantraniliprole	solvent	1–100	$Y = 3.48823 \times 10^7X + 43{,}168.4$	0.9958	
	litchi	1–100	$Y = 1.45219 \times 10^7X + 190{,}031$	0.999	−58.41
	longan	1–100	$Y = 2.00116 \times 10^7X + 14{,}557.3$	0.9954	−42.6
Cyantraniliprole	solvent	1–100	$Y = 7.59264 \times 10^6X - 4650.42$	0.9998	
	litchi	1–100	$Y = 2.38347 \times 10^6X + 685.684$	0.9989	−65.42
	longan	1–100	$Y = 3.98107 \times 10^6X - 3974.61$	0.9962	−43.01
IN-J9Z38	solvent	1–100	$Y = 2.41814 \times 10^6X - 3692.34$	0.9972	
	litchi	1–100	$Y = 3.61443 \times 10^5X + 908.734$	0.992	−79.12
	longan	1–100	$Y = 1.62265 \times 10^6X + 3504.66$	0.9917	−33.51
Imidacloprid	solvent	1–100	$Y = 2.19412 \times 10^7X + 11{,}740.08$	0.9962	
	litchi	1–100	$Y = 6.56189 \times 10^6X - 1297.32$	0.9989	−70.14
	longan	1–100	$Y = 5.0 \times 10^6X + 68{,}417$	0.999	−77.19

[a] ME means matrix effect, and was calculated with the equation: ME = (slope of the matrix-matched standard/slope of the solvent standard − 1) × 100%. An ME with a negative and positive value represents that the pesticide response is suppressed and enhanced. It is generally believed that when: $|ME| < 20\%$ the matrix does not exist; $20\% \leq |ME| \leq 50\%$ it indicates a medium matrix effect; and when $|ME| > 50\%$ it indicates a strong matrix effect.

3.9. Method Validation

The recovery rates, RSDs, limit of detection (LODs), and LOQs of the 7 pesticides and a metabolite residue in litchi and longan are shown in Table 4. Good linearity was acquired for correlation coefficient values of >0.99. The average recovery rate was 81–99% in matrices, with an RSD of 3.5–8.4%, which represent satisfactory precision and accuracy. These results were compliant with the rules stating that the mean recovery rate should be in the range of 70–120% with an associated RSD of ≤20%. The LOQ was 1–10 μg/kg for both litchi and longan. The LOD was the concentration that produced a signal-to-noise ratio of 3 and was 0.3–3 μg/kg.

3.10. Application in Real Samples

Ten litchi and longan samples randomly purchased from the market were examined using the validated UPLC-MS/MS method for monitoring the aforementioned 8 target analytes. Azoxystrobin, pyraclostrobin, difenoconazole, and chlorantraniliprole were detected. Among them, the MRL value of azoxystrobin, pyraclostrobin, and difenoconazole were established in litchi [42]. The residues in the real samples did not exceed the regulated MRLs (Table 5).

Table 4. Recoveries, RSDs, LOD, and LOQ of eight targeted compounds in litchi and longan samples (n = 5).

Compound	Sample	Spiked Level /(μg/kg)	Average Recoveries [a] /%	Relative Deviation [b] /%	LOD /(μg/kg)	LOQ /(μg/kg)
Azoxystrobin	litchi	1, 10, 100	83, 94, 94	3.9, 7.5, 6.9	0.3	1
	longan	1, 10, 100	81, 89, 94	3.5, 6.8, 8.2	0.3	1
Pyraclostrobin	litchi	1, 10, 100	86, 94, 95	7.1, 8.2, 8.1	0.3	1
	longan	1, 10, 100	85, 96, 94	6.0, 7.5, 4.0	0.3	1
Picoxystrobin	litchi	1, 10, 100	84, 94, 95	4.4, 6.6, 7.4	0.3	1
	longan	1, 10, 100	86, 92, 96	7.1, 8.4, 4.7	0.3	1
Difenoconazole	litchi	1, 10, 100	84, 91, 94	4.1, 6.1, 5.0	0.3	1
	longan	1, 10, 100	90, 97, 96	5.0, 4.1 5.4	0.3	1
Chlorantraniliprole	litchi	1, 10, 100	85, 93, 93	5.8, 6.3, 4.7	0.3	1
	longan	1, 10, 100	82, 95, 94	4.3, 6.1, 7.0	0.3	1
Cyantraniliprole	litchi	1, 10, 100	85, 96, 99	4.7, 6.1, 4.7	0.3	1
	longan	1, 10, 100	84, 91, 99	5.2, 6.7, 5.6	0.3	1
IN-J9Z38	litchi	1, 10, 100	85, 89, 95	4.3, 5.6, 7.7	3	10
	longan	1, 10, 100	86, 96, 98	5.2, 7.5, 4.7	3	10
Imidacloprid	litchi	1, 10, 100	83, 97, 97	4.5, 7.0, 5.1	0.3	1
	longan	1, 10, 100	84, 95, 98	5.3, 6.3, 5.4	0.3	1

[a] The recovery was calculated by the formula: Recovery = C_d /C_s × 100%, where C_d represents the detected concentration and C_s represents the spiked concentration. Results were expressed as mean ± standard deviation (SD) with 95% confidence intervals. [b] Mean value of five determinations.

Table 5. Residues of eight targeted compounds in real litchi and longan samples.

Matrix MRLs	Azoxystrobin	Pyraclostrobin	Picoxystrobin	Difenoconazole	Chlorantraniliprole	Cyantraniliprole	IN-J9Z38	Imidacloprid
				mg/kg				
Litchi	ND-0.159	ND-0.091	ND	ND-0.458	ND-0.019	ND	ND	ND
Longan	ND-0.17	ND-0.08	ND	0.02–0.16	0.02–0.03	ND	ND	ND
MRLs in litchi	0.5	0.1	-	0.5	-	-	-	-
MRLs in longan	-	-	-	-	-	-	-	-

Note: ND is none of detection.

4. Conclusions

We simultaneously determined and analyzed azoxystrobin, pyraclostrobin, picoxystrobin, difenoconazole, chlorantraniliprole, imidacloprid, and cyantraniliprole, and a cyantraniliprole metabolite IN-J9Z38 in litchi and longan by using optimized sample pretreatment, instrument conditions, and QuEChERS–HPLC-MS/MS. The samples were homogeneously extracted with acetonitrile, purified using the improved QuEChERS method, and detected through HPLC-MS/MS. The Infinity Lab Poroshell 120 SB-C18 chromatographic column, electrospray ionization, positive ion scanning, and MRM were used for separation, analysis, detection, and quantification of the target analytes. The matrix matching standard solution was determined using the external standard method. The linear relationship of the 8 target compounds within the range of 1–100 μg/L was good, and the correlation coefficients were >0.99. At the spiked level of 1, 10, and 100 μg/kg, the average recovery of eight target compounds was 81–99% and the RSD was 3.5–8.4%. The pretreatment process of the method is simple and rapid, and the detection limit, precision, and linear range of the method could meet the requirements. This method can be used for the simultaneous determination and analysis of the aforementioned pesticide residues in litchi and longan samples.

Author Contributions: Methodology, data curation, writing-original draft preparation, S.W.; validation, data curation, X.W.; sample analysis, and data curation, Q.H. and H.L.; literature search, figures, H.C.; funding acquisition, sampling, H.S.; supervision, data analysis, Y.L. All authors have read and agreed to the published version of the manuscript.

Funding: This research was funded by Guangzhou Science and Technology grant number [202102080338]; the Special Fund for Rural Revitalization Strategy of Guangdong province grant number [403-2018-XMZC-0002-90].

Institutional Review Board Statement: Not applicable.

Informed Consent Statement: Not applicable.

Data Availability Statement: Not applicable.

Acknowledgments: This work was supported by earmarked fund for CARS-32 (Project No. CARS-32-13), and Guangzhou Science and Technology project (Project No. 202201010479).

Conflicts of Interest: The authors declare no conflict of interest.

References

1. Sun, H.Y.; Hu, X.C.; Zhang, H.J. Analysis of distribution of litchi-related literature based on SCI-E. *Chin. J. Trop. Agric.* **2018**, *38*, 123–130.
2. Bai, H.Q.; Wu, J.G.; Pan, X.B. Key climatic factors affecting the distribution of litchi in China. *J. Fruit Sci.* **2016**, *33*, 436–443.
3. Qi, W.E.; Chen, H.B.; Li, W.W.; Zhang, H.J. Development situation, trend and suggestions of Chinese litchi industry. *Guangdong Agric. Sci.* **2016**, *43*, 173–179.
4. Hao, J.; Dong, L.H.; Chi, J.W.; Ma, Y.X.; Zhou, Q.Y.; Zhang, R.F.; Bai, Y.J. Health Effect of Longan and Its Present Situation and Prospect of Healthy. *Mod. Food Sci. Technol.* **2021**, *37*, 340–349.
5. Zhao, F.; Liao, M.J.; Zhang, J.E.; Huang, M.; Zhang, Z.H. Characteristics, value, and conservation of litchi heritage systems in China: A case study of the Lingnan Litchi Cultivation System (Zengcheng). *Chin. J. Eco-Agric.* **2020**, *28*, 1435–1442.
6. Ma, R.H.; Xie, X.M.; Xiao, R.Y.; Lei, S.M.; Han, G.; Lin, J.; Yang, Z.Y.; Song, M.H.; Jiang, M.; Zeng, W.Y.; et al. Study of fruit quality of main longan cultivars in Chognqing. *J. Southwest Univ.* **2021**, *43*, 52–60.
7. Li, P.; Huo, L.; Su, W.; Lu, R.M.; Deng, C.C.; Liu, L.Q.; Deng, Y.K.; Guo, N.N.; Lu, C.S.; He, C.L. Free radical-scavenging capacity, antioxidant activity and phenolic content of Pouzolziazeylanica. *J. Serb. Chem. Soc.* **2011**, *76*, 709–717. [CrossRef]
8. Du, Q.; Xu, Y.; Li, L.; Zhao, Y.; Gerold, J.; Peter, W. Antioxidant constituents in the fruits of Luffa cylindrica (L.) Roem. *J. Agric. Food Chem.* **2006**, *54*, 4186–4190. [CrossRef]
9. China Pesticide Information Network. Available online: http://www.chinapesticide.org.cn/hysj/index.jhtml. (accessed on 18 August 2021).
10. Liu, Y.P.; Wang, S.W.; Sun, H.B. Present situation of pesticide residues in litchi and analysis of formulation of maximum residue limit in China. *China Trop. Agric.* **2016**, *47*, 31–34.
11. Whitehorn, P.R.; O'connor, S.; Wackers, F.L.; Wackers, F.L.; Goulson, D. Neonicotinoid pesticide reduces bumble bee colony growth and queen production. *Science* **2012**, *336*, 351–352. [CrossRef]
12. Rundlof, M.; Andersson, G.S.; Bommarco, R.; Fries, I.; Hederstrom, V.; Herbertsson, L.; Jonsson, O.; Klatt, B.K.; Pedersen, T.R.; Yourstone, J.; et al. Seed coating with aneonicotinoid insecticide negatively affects wild bees. *Nature* **2015**, *521*, 77–80. [CrossRef]
13. Suchail, S.; Guezd, L.; Belzunces, L.P. Discrepancy between acute and chronic toxicity induced by imidacloprid and its metabolites in Apismellifera. *Environ. Toxicol. Chem.* **2001**, *20*, 2482–2486. [CrossRef]
14. Park, M.G.; Blitzer, E.J.; Gibbs, J.; Losey, J.; Danforth, B.N. Negative effects of pesticides on wild bee communities can be buffered by landscape context. *Proc. R. Soc. B* **2015**, *282*, 20150299. [CrossRef]
15. Yoder, J.A.; Jajack, A.J.; Rosselot, A.E.; Smith, T.J.; Yerke, M.C.; Sammataro, D. Fungicide contamination reduces beneficial fungi in bee bread based on an area-wide field study in honey bee, Apis mellifera, colonies. *J. Toxicol. Environ. Health A* **2013**, *76*, 587–600. [CrossRef]
16. Bernauer, O.M.; Gaines-day, H.R.; Steffan, S.A. Colonies of bumble bees (bombus impatiens) produce fewer workers, less bee biomass, and have smaller mother queens following fungicide exposure. *Insects* **2015**, *6*, 478–488. [CrossRef]
17. Crowell, S.R.; Henderson, W.M.; Kenneke, J.F.; Fisher, J.W. Development and application of a physiologically based pharmacokinetic model for triadimefon and its metabolite triadimenol in rats and humans. *Toxicol. Lett.* **2011**, *205*, 154–162. [CrossRef] [PubMed]
18. Goetz, A.K.; Dix, D.J. Mode of action for reproductive and hepatic toxicity inferred from a genomic study of triazole antifungals. *Toxicol. Sci.* **2009**, *110*, 449–462. [CrossRef] [PubMed]
19. Liu, Y.P.; Wang, S.W.; Wang, X.N.; Chang, H.; Sun, H.B. Residue of pyraclostrobin and its metabolite BF-500-3 in Dimocarpus longan Lour. *Chin. J. Pestic. Sci.* **2020**, *22*, 864–868.
20. Sun, C.X.; Ouyang, Z.Z.; Liu, Y.H.; Yu, G.G. Residue dynamic and risk assessment of three fungicides in.broccoli. *Acta Agric. Zhejiangensis* **2021**, *33*, 1292–1299.
21. Xu, X.M.; Long, Z.Y.; Huang, W.X.; Yue, Y.J.; Lv, F.; Huang, X.D.; Liu, R. Simultaneous determination for 35 kinds of strobilurin fungicides and triazole fungicide in fruit and vegetable by QuEChERS-ultra performance liquid chromatography-tandem mass spectrometry. *J. Food Saf. Qual.* **2019**, *10*, 1039–1047.

22. Mao, J.S.; Chen, z.l.; Li, H.D.; Zhang, W.J.; Ding, R.Y.; Fang, L.P.; Guo, C.Y. Residues and dissipation dynamics of chlorpyrifos, imidacloprid, spirotetramat and its motablItes, difenoconazole in pear. *Chin. J. Pestic. Sci.* **2019**, *21*, 395–400.
23. Liang, X.M.; Zhang, W.Y.; Zhang, W.; Xie, S.B.; He, R.Y.; Xu, J.; Su, L.J.; Sun, B.L.; Zhang, J. Simultaneous determination of residues of 38 pesticides in fruits by QuEChERS combined with high performance liquid chromatography-tandem mass spectrometry. *Food Sci.* **2020**, *41*, 288–296.
24. Wu, C.X.; Zhang, Q.P.; Xu, H.R.; Wang, C.M. Rapid determination of 4 pesticide residues in vegetables by QuEChERS and ultra performance liquid chromatography tandem mass spectrometry. *J. Food Saf. Qual.* **2020**, *11*, 909–914.
25. Yang, S.; Wang, Y.; He, L.F.; Huang, X.P.; Mu, W. Residue and dissipation dynamics of six insecticides in fresh tea by QuEChERS-gas chromatography tandem mass spectrometry. *Chin. J. Anal. Lab.* **2019**, *38*, 1459–1464.
26. Dong, H.; Xian, Y.P.; Li, H.X.; Wu, Y.L.; Bai, W.D.; Zeng, X.F. Analysis of heterocyclic aromatic amine profiles in Chinese traditional bacon and sausage based on ultrahigh-performance liquid chromatograph-quadrupole-orbitrap high-resolution mass spectrometry (UHPLC-Q-Orbitrap-HRMS). *Food Chem.* **2020**, *310*, 125937. [CrossRef] [PubMed]
27. Gago-Ferrero, P.; Bletsou, A.A.; Damalas, D.E.; Aalizadeh, R.; Alygizakis, N.A.; Singer, H.P.; Hollender, J.; Thomaidis, N.S. Wide-scope target screening of & 2000 emerging contaminants in wastewater samples with UPLC-Q-ToF-HRMS/MS and smart evaluation of its performance through the validation of 195 selected representative analytes. *J. Hazard. Mater.* **2020**, *387*, 121712. [PubMed]
28. Lu, Z.; Zhang, Z.B.; Fang, N.; Hou, Z.G.; Li, Y.R.; Lu, Z.B. Simultaneous determination of five diamide insecticides in food matrices using carbon nanotube multiplug filtration cleanup and ultra-high-performance liquid chromatography-tandem mass spectrometry. *J. Agric. Food Chem.* **2019**, *67*, 10977–10983. [CrossRef]
29. Dong, M.F.; Ma, L.; Zhan, X.P.; Chen, J.B.; Huang, L.Q.; Wang, W.M.; Zhao, L. Dissipation rates and residue levels of diflubenzuron and difenoconazole on peaches and dietary risk assessment. *Regul. Toxicol. Pharmacol.* **2019**, *18*, 104447–104450. [CrossRef]
30. Abdallah, O.; Arasheed, A.; Almundarij, A.A.; Omar, A.F.; Alhewairini, S.S.; Aljamhanm, K.A. Levels of residues and dietary risk assessment of the fungicides myclobutanil, penconazole, tebuconazole and triadimenol in squash. *Biomed. Chromatogr.* **2021**, *35*, e5126. [CrossRef]
31. Ma, S.P.; Yuan, X.C.; Zhao, P.F.; Sun, H.; Ye, X.; Liang, N.; Zhao, L.S. Trace determination of five triazole fungicide residues in traditional Chinese medicine samples by dispersive solid-phase extraction combined with ultrasound-assisted dispersive liquid–liquid microextraction and UHPLC–MS/MS. *J. Sep. Sci.* **2017**, *40*, 3257–3266. [CrossRef]
32. Zhang, Y.H.; Zhang, Y.; Nie, J.Y.; Jiao, B.N.; Zhao, Q.Y. Determination of Triazole Fungicide Residues in fruits by QuEChERS combined with ionic liquid-Based dispersive liquid-liquid microextraction: Optimization using response surface methodology. *Food Anal. Methods* **2016**, *9*, 3509–3519. [CrossRef]
33. Luo, J.; Bian, C.F.; Rao, L.; Zhou, W.W.; Li, Y.Q.; Li, B.T. Determination of the residue behavior of.isocycloseram in Brassica and soil using the QuEChERS method coupled with HPLC. *Food Chem.* **2022**, *367*, 130734. [CrossRef] [PubMed]
34. Li, L.C.; Yin, Y.; Zheng, G.M.; Liu, S.G.; Zhao, C.; Xie, W.P.; Ma, L.S.; Shan, Q.; Dai, X.X.; Wei, L.T. Determination of multiclass herbicides in sediments and aquatic products using QuECHERS combined with ultra-high performance liquid chromatography-tandem mass spectrometry (UHPLC-MS/MS) and its application to risk assessment of rice-fish co-culture system in China. *Microchem. J.* **2021**, *170*, 106628.
35. Ji, B.C.; Zhao, W.H.; Xu, X.; Han, Y.; Jie, M.S.; Xu, G.G.; Bai, Y.H. Development of a modified quick, easy,.cheap, effective, rugged, and safe method based on melamine sponge for multi-residue analysis of veterinary drugs in milks by ultra-performance liquid chromatography tandem mass spectrometry. *J. Chromatogr. A* **2021**, *1651*, 462333. [CrossRef]
36. Ferreira, J.A.; Ferreira, J.M.S.; Talamini, V.; Facco, J.F.; Rizzetti, T.M.M.; Prestes, O.D.; Adaime, M.B.; Zanella, R.; Bottoli, C.B.G. Determination of pesticides in coconut (Cocos nucifera Linn.) water and pulp using modified QuEChERS and LC-MS/MS. *Food Chem.* **2016**, *213*, 616–624. [CrossRef]
37. Sapozhnikova, Y.; Lehotay, S.J. Multi-class, multi-residue analysis of pesticides, polychlorinated biphenyls, polycyclic aromatic hydrocarbons, polybrominated diphenyl ethers and novel flame retardants in fish using fast, low-pressure gas chromatography–tandem mass spectrometry. *Anal. Chim. Acta.* **2013**, *758*, 80–92. [CrossRef]
38. Erney, D.R.; Gillespie, A.M.; Gilvydis, D.M.; Poole, C.F. Explanation of the matrix-induced chromatographic response enhancement of organophosphorus pesticides during open tubular column gas chromatography with splitless or hot on-column injection and flame photometric detection. *J. Chromatogr. A* **1993**, *638*, 57–63. [CrossRef]
39. Tan, J.L.; Peng, Z.H.; Zhao, X.L.; Ma, X.T.; Zhang, X.Y.; Liu, X.L.; Hu, Z.B.; Feng, L. Determination of 19 carbamate pesticide and metabolites residues in eggs by QuEChERS-UP-C-MS/MS. *Sci. Technol. Food Ind.* **2022**, *43*, 320–325.
40. Kmellar, B.; Fodor, P.; Pareja, L.; Ferrer, C.; Martinez-Uroz, M.A.; Valverde, A.; Fernandez-Alba, A.R. Validation and uncertainty study of a comprehensive list of 160 pesticide residues in multi-class vegetables by liquid chromatography-tandem mass spectrometry. *J. Chromatogr. A* **2008**, *1215*, 37–50. [CrossRef]
41. Bin, J.; Wang, Y.X.; Hu, J.L.; Feng, L.; Tan, J.L. Rapid determination of 33 pesticide residues in Xiangmao grass by QuEChERS UPLC MS/MS. *Chin. J. Pharm. Anal.* **2021**, *41*, 471–479.
42. *GB 2763—2021*; National Food Safety Standard-Maximum Residue Limits for Pesticides in Food. China Standard Press: Beijing, China, 2021.

molecules

MDPI

Article

Evaluation of the Antioxidant Capacity of Fruit Juices by Two Original Analytical Methods

Michele Protti [1], Isacco Gualandi [2], Sergio Zappoli [2], Roberto Mandrioli [3], Laura Mercolini [1] and Domenica Tonelli [2,*]

[1] Department of Pharmacy and Biotechnology (FaBiT), Alma Mater Studiorum-University of Bologna, Via Belmeloro 6, 40126 Bologna, Italy; michele.protti2@unibo.it (M.P.); laura.mercolini@unibo.it (L.M.)
[2] Department of Industrial Chemistry "Toso Montanari", Alma Mater Studiorum-University of Bologna, Viale del Risorgimento 4, 40136 Bologna, Italy; isacco.gualandi2@unibo.it (I.G.); sergio.zappoli@unibo.it (S.Z.)
[3] Department for Life Quality Studies (QuVi), Alma Mater Studiorum-University of Bologna, Corso d'Augusto 237, 47921 Rimini, Italy; roberto.mandrioli@unibo.it
* Correspondence: domenica.tonelli@unibo.it; Tel.: +39-051-209-3667

Abstract: Two analytical methods previously developed by our groups were employed to estimate the antioxidant capacity of commercial fruit juices. The electrochemical method, which measures the scavenging activity of antioxidants towards OH radicals generated by both hydrogen peroxide photolysis and Fenton's reaction, is based on the recovery of the cyclic voltametric response of the redox probe $Ru(NH_3)_6^{3+}$ at a Glassy Carbon electrode modified with a thin film of an insulating polyphenol, in the presence of compounds with antioxidant properties. The values of the antioxidant capacity of the fruit juices are expressed as vitamin C equivalents/L. The chromatographic method is based on the generation of OH radicals via Fenton's reaction in order to test the inhibition of their formation in the presence of antioxidant compounds by monitoring salicylate aromatic hydroxylation derivatives as markers of •OH production, by means of HPLC coupled to coulometric detection. The results are expressed as the percentage of inhibition of •OH production in the presence of the tested juice compared to the control sample. When OH radicals are produced by Fenton's reaction, the antioxidant capacity of the juices, estimated by both methods, displays an analogous trend, confirming that they can be considered an alternative for measuring the ability of antioxidants to block OH radical formation.

Keywords: antioxidant capacity; hydroxyl radicals; Fenton's reaction; hydrogen peroxide photolysis; HPLC with coulometric detection; polyphenol modified electrode

Citation: Protti, M.; Gualandi, I.; Zappoli, S.; Mandrioli, R.; Mercolini, L.; Tonelli, D. Evaluation of the Antioxidant Capacity of Fruit Juices by Two Original Analytical Methods. *Molecules* **2023**, *28*, 6672. https://doi.org/10.3390/molecules28186672

Academic Editor: Gavino Sanna

Received: 14 July 2023
Revised: 15 September 2023
Accepted: 16 September 2023
Published: 18 September 2023

1. Introduction

The recent shift from the traditional medical care approach to personalized medicine, which emphasizes the prevention rather than the cure of diseases, has raised awareness of how important lifestyles and the quality of nutrition are. In this context, antioxidants from fruits and vegetables are considered bioactive dietary compounds which can reduce oxidative stress, and epidemiological data have indicated an inversed correlation between the intake of fruits and vegetables, which are naturally rich in antioxidants, and the incidence of some illnesses such as cardiovascular or metabolic disorders and cancer [1]. Oxidative stress takes place when an unbalance occurs between chemically reactive oxidative species, i.e., reactive oxygen species (ROS), reactive nitrogen species (RNS), reactive sulfur species (RSS) and reactive carbonyl species (RCS), and the ability of the organism to counteract their action through its antioxidant protective systems, including both antioxidant enzymes and endogenous antioxidants [2].

From a nutritional point of view, an antioxidant is defined as any compound which, when present in low concentrations as compared to those of an oxidable substrate, is able to significantly delay or inhibit the oxidation of that substrate [2].

Therefore, the analysis of these compounds and the evaluation of the antioxidant activity/capacity of various foods and beverages have grown considerably in recent years [3]. Unfortunately, a common issue in this research field is how to measure such an activity/capacity and express the results. In fact, although a great effort has been made to standardize the analytical methods, their results are often lacking concordance, and that can be ascribed to the different mechanisms that antioxidants play and the different principles on which the various assays for measuring antioxidant capacity are based [4].

Among the most commonly employed assays for evaluating the antioxidant properties of food, we can cite DPPH, ABTS, Folin-Ciocalteau, CUPRAC, FRAP, ORAC and TRAP [4].

In 2015, Tonelli's group proposed an electrochemical method (OH-RSC), which measured the scavenging activity of antioxidants towards OH radicals generated by hydrogen peroxide photolysis. The method was based on the recording of the cyclic voltametric response (CV) of a redox probe, $Ru(NH_3)_6{}^{3+}$, at a Glassy Carbon electrode (GCE) modified with a thin film of an insulating polyphenol, after it had been exposed to OH radicals. In fact, the intensity of the recovered electrochemical signal was proportional to the extent of the polymer degradation, as the redox probe was fully unresponsive when the polyphenol film was just electrodeposited on the GCE. The antioxidant capacity (AOC) of some standard compounds and fruit juices estimated by the OH-RSC method was compared with that obtained using more standardized assays, like ABTS, DPPH and ORAC. The best correlation among the obtained results was noticed with the ORAC assay [5,6]. Later, the electrochemical method was applied to the analysis of the AOC of dried goji berries [7], which were also analyzed by an LC–MS/MS method to determine the concentration of 23 different analytes with well-known antioxidant properties. In that work, an extraction and a pretreatment procedure were also developed, and the overall analytical method was validated according to international guidelines. The conclusion was that there was a substantial agreement between the results obtained from the two investigations.

Mercolini's group has recently developed and optimized an in vitro Fenton's system to test the variation of •OH formation by monitoring salicylate aromatic hydroxylation derivatives as markers of •OH production, namely, 2,3- and 2,5-dihydroxybenzoic acid (2,3-DHBA and 2,5-DHBA) and catechol. Salicylic acid was chosen as the substrate because it easily undergoes radical hydroxylation, producing catechol, which originates from the aromatic radical addition of a hydroxyl group in position one, followed by decarboxylation, while the other two products are derived by the same addition in other activated positions (three and five). These molecules are electroactive due to the vicinal phenol groups (catechol) that can be easily oxidized or reduced [8], and the products can be detected by electrochemical detection (ED) with high sensitivity, particularly by coulometric detection thanks to the specific cell geometry allowing for consistent and extensive compound oxidation (or reduction) [9]. The target analytes were determined by HPLC coupled to coulometric detection using a fully validated method, which was then exploited to evaluate the antioxidant and/or prooxidant activity of various bioactive molecules by comparing hydroxylation products' (and, thus, •OH) formation rates in Fenton's reaction mixtures with and without bioactive molecules [10].

The aim of the present work was to analyze samples of fruit juices commercially available in the Italian market by means of the analytical methods proposed by Tonelli's and Mercolini's groups [6,9] to verify if they can be considered in some way equivalent in estimating the antioxidant capacity, thus providing an alternative yet reliable platform to be exploited in investigating both the antioxidant and prooxidant capacity of significant bioactive compounds, natural extracts and foods rich in chemically different compounds, also paving the way for more detailed studies on the antioxidant and prooxidant mechanism of action of in vivo complex systems.

2. Results and Discussion

In Table 1, the results obtained from the chromatographic method when applied to the reference standard Fenton's reaction in terms of the mean peak area of the three hydroxylation products are reported.

Table 1. Method application to a reference standard Fenton's reaction [a].

Analyte	Average Concentration (nM)	RSD%
Catechol	201	2
2,5-DHBA	151	3
2,3-DHBA	384	2
Sum	7.4×10^2	2

[a] n = 5.

Figure 1 reports typical HPLC-ED chromatograms obtained from the analysis of a reference Fenton's reaction mixture alone and in the presence of a fruit juice sample.

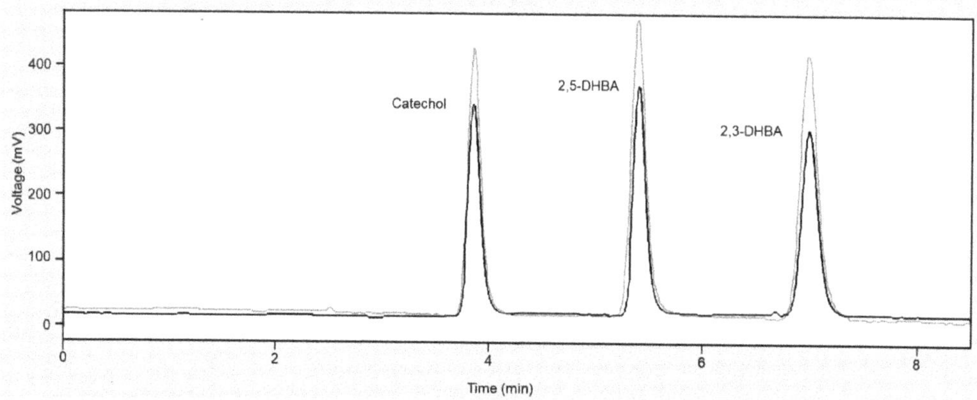

Figure 1. HPLC-ED chromatograms obtained from the analysis of salicylic acid hydroxylation products after carrying out a standard Fenton's reaction (grey line) and in the presence of a fruit juice sample (black line).

In Table 2, the data obtained from the two analytical methods for all the analyzed juices are shown. As already stated above, the results from the electrochemical method are expressed as Vitamin C equivalents/L of the sample, whereas the results from the chromatographic method are expressed as the percentage of inhibition of •OH production in comparison to that of the control sample not containing the tested juice.

In fact, in such a case, a decrease in the production of catechol, 2,3-DHBA and 2,5-DHBA, as evidenced by the chromatograms in Figure 1, indicates a corresponding proportional decrease in •OH production, which is consequently related to the antioxidant capacity of the fruit juice under investigation.

Even if the results from the two methods are calculated in a different way, it is easy to draw some conclusions.

Taking into account the data coming from the electrochemical method, it is possible to say that the VCE values are significantly higher when the OH radicals are produced by the Fenton's reaction than when the H_2O_2 photolysis is employed. That is explainable considering that, in the former case, the modified electrode estimates the ability of antioxidants to block OH radical formation, whereas in the latter, it estimates the ability to remove OH radicals. Moreover, the VCE indexes of the different juices from the photolytic production of OH radicals are quite close together since the maximum value (pomegranate juice, sample J) is 1.68 and the minimum (blueberries mix juice, sample F) is 0.25 mol/L, respectively. On the contrary, the data from the Fenton's reaction allow for highlighting a net difference between some samples that result in possessing a very high AOC (samples K and J, corresponding to a goji berry and pomegranate juice, respectively, with values of 22.7 and 16.9 mol/L) and those that result in exhibiting a very low AOC (samples C and A, with a VCE less than 1). In addition, a very good reproducibility for the analytical measurement of real samples AOC was observed when the OH radicals were produced by Fenton's reaction, as highlighted by the percentage RSD values.

Table 2. Method application to fruit juice samples.

Sample	Description	Analyte Concentration Sum [a] (μM)	Variation with Respect to Standard Fenton [a] $((C_{sample} - C_{Fenton})/C_{Fenton})$	Vitamin C Equivalents OH-RSC [a] (mol/L) (Fenton)	Vitamin C Equivalents OH-RSC [a] (mol/L) (Photolysis)
A	Blue fruits mix	0.624 (1)	−0.15 (1 × 10)	0.70 (4)	0.26 (2 × 10)
B	Yellow fruits mix	0.62 (2)	−0.15 (1 × 10)	1.74 (1)	0.7 (1 × 10)
C	Violet fruits mix	0.62 (3)	−0.16 (2 × 10)	0.94 (2)	0.85 (7)
D	Berries and red fruits	0.60 (7)	−0.18 (3 × 10)	1.29 (1)	0.68 (9)
E	Red fruits mix	0.60 (5)	−0.19 (3 × 10)	2.40 (1)	1.02 (6)
F	Blueberries mix	0.63 (2)	−0.15 (1 × 10)	1.05 (2)	0.25 (2 × 10)
G	Red fruits mix	0.62 (2)	−0.15 (1 × 10)	1.49 (1)	0.31 (2 × 10)
H	Carrot bio-centrifuged juice	0.60 (5)	−0.18 (3 × 10)	2.48 (1)	0.40 (1 × 10)
I	Cranberry juice	0.61 (3)	−0.17 (2 × 10)	1.80 (1)	1.33 (4)
J	Pomegranate juice	0.56 (4)	−0.23 (2 × 10)	16.9 (1)	1.68 (2)
K	Goji juice and pulp	0.58 (5)	−0.21 (2 × 10)	22.7 (1)	0.65 (6)

[a] Average (percentage relative standard deviation, RSD%), n = 5.

By comparing the results coming from the two analytical methods, considering again the Fenton's reaction as the source of OH radicals for the electrochemical one, it is possible to observe that three groups of samples can be identified; the first consists of two juices which possess a very high AOC, the second consists of four juices whose AOC can be considered intermediate and the third consists of five juices which display a lower AOC.

In particular, the juices belonging to the three groups are the same for both methods, with only the exception of samples B and D, which are exchanged between the second and the third group. Therefore, we can state that the trend of the AOC values for the tested samples is in good agreement when both the proposed methods are employed. To further corroborate this conclusion, the AOC of the samples was also determined by three analytical methods, commonly employed for food analysis, i.e., ABTS, DPPH and ORAC, according to an experimental procedure already published [5]. The relevant results are shown in Table 3. In such a case, they are expressed as mmol Vitamin C equivalents per L of the fruit juice, since ABTS, DPPH and ORAC generally provide lower values than the ones obtained from the electrochemical method, and, in particular, those from the DPPH assay are the lowest, as already pointed out [5], as it is based on the scavenging capacity of antioxidants towards a rather stable nitrogen radical.

Table 3. Application of standard methods to fruit juice samples.

Sample	Vitamin C Equivalents ABTS [a] (mmol/L)	Vitamin C Equivalents DPPH [a] (mmol/L)	Vitamin C Equivalents (ORAC) [a] mmol/L
A	10.9 (6)	1.63 (6)	80.7 (1)
B	8.70 (6)	1.20 (8)	63.7 (1)
C	5.30 (1 × 10)	1.14 (9)	93.6 (1)
D	4.32 (1 × 10)	3.15 (6)	170.1 (1)
E	6.24 (8)	2.45 (8)	169.9 (1)
F	5.26 (9)	2.35 (4)	66.0 (1)
G	6.33 (8)	2.20 (4)	77.0 (1)
H	12.90 (3)	4.15 (9)	105.5 (1)
I	9.30 (5)	1.81 (6)	77.7 (1)
J	26.90 (2)	4.30 (7)	110.5 (1)
K	20.90 (2)	3.58 (8)	89.3 (1)

[a] Average (RSD%), n = 5.

Although the numerical values are different and depend on the employed analytical method, it is possible to highlight, again, three groups of samples displaying high, intermediate and low AOC values. In any case, the sample J is the most active, and the sample F is one of the least active, when analyzed by ORAC, DPPH and ABTS assays, and this result is in rather good agreement with the data displayed in Table 2, relevant to the two methods proposed here for the evaluation of the AOC of fruit juices. In conclusion, even if the number of analyzed samples is not particularly high, the electrochemical and chromatographic methods can be considered alternative approaches to estimate the antioxidant capacity of real samples since both measure the ability of antioxidants to block OH radical formation. This result is remarkable, as the OH radical formation and inhibition in the chromatographic method always occur in a solution, whereas in the electrochemical method, the OH radicals must arrive from the solution to the modified electrode and attack the polyphenol film in order to be quantified. In the latter case, it is more likely that other compounds present in the fruit juices can react with the OH radicals, and this fact can explain why the VCEs of the samples with the highest (K) and lowest value (A) are in a ratio greater than 30, whereas the inhibition percentage ratio between the sample displaying the highest inhibition and that with the lowest is about 1.5.

3. Materials and Methods

3.1. Reagents

All chemicals and reagents were of analytical grade and used as received. L-Ascorbic acid (vitamin C), hexa-aminoruthenium(III) chloride, hydrogen peroxide solution (30% *w*/*w*), $FeSO_4$, catechol, 2,3-DHBA, 2,5-DHBA, salicylic acid, CuCl, NaCl, $NaHCO_3$, H_3PO_4, triethylamine and HPLC-grade acetonitrile were bought from Merck Life Science. CH_3COONa, phenol and NaOH were purchased from Carlo Erba (Milan, Italy). Acetic acid and sulfuric acid were bought from Baker. Ultrapure water, obtained from a Millipore Milli-Q system (Burlington, MA, USA), was used throughout the experiments.

3.2. Electrochemical Method

The experimental procedure for evaluating the AOC has been already published [5]. Here, it is only briefly summarized.

The GCE was cleaned until a mirror-like surface was obtained; then, it was modified with the polyphenol film by performing a potentiostatic electropolymerization of phenol (0.05 M) in 1 M sulfuric acid, applying a potential of +1.0 V for 60 s. An aqueous saturated calomel electrode and a Pt wire were used as the reference and counter electrode, respectively. To check if a well adhered film had been deposited, a CV in 0.5 M acetate buffer, pH 4.6, containing the redox probe $Ru(NH_3)_6^{3+}$ (5×10^{-3} M), was recorded (blank signal). The absence of the redox probe signal indicated a satisfactory deposition.

The AOC of fruit juices was evaluated by studying the degradation kinetics of the polyphenol film in the absence and in the presence of the food samples, generating the OH radicals either by hydrogen peroxide photolysis or by Fenton's reaction triggered by Fe(II) ions. Depending on the type and concentration of the antioxidants present in the fruit juice, the time requested to induce the polymer degradation increases due to the protective action of such compounds.

The photolytic generation of OH radicals was performed using a 120 mL reactor containing a 0.01 M H_2O_2 solution, where the modified electrode was placed near the lamp operating at 254 nm (UV18F, Italquarz, with emission power of 17 W) using a homemade holder that ensured the reproducibility of the sensor position during irradiation. In order to obtain the calibration line for vitamin C and to estimate the AOC of the analyzed juices, proper amounts of the standard antioxidant or of the samples were added. After irradiation, the electrode was taken out from the reactor at prefixed times, rinsed with ultrapure water and put in acetate buffer, where the CV signal related to the $Ru(NH_3)_6^{3+}$ redox system was again recorded. The total irradiation time was generally 30 min.

For the generation of OH radicals by Fenton's reaction, 20 mL of a 0.05 M $FeSO_4$ solution containing a proper amount of Vitamin C or the sample were poured in an electrochemical cell where the modified GCE had been inserted. In total, 800 µL of 10 M H_2O_2 solution were added to start the reaction. After 10 min, the electrode was taken out from the solution, rinsed with ultrapure water and put in acetate buffer, where the CV signal related to the $Ru(NH_3)_6^{3+}$ redox system was recorded.

3.3. *Chromatographic Method*

3.3.1. Fenton Reaction

The formation of •OH via Fenton chemistry was monitored by the quantitation of three salicylic acid oxidation products: catechol, 2,3-DHBA and 2,5-DHBA.

A standard Fenton's reaction in the presence of copper was triggered by adding 7.5 µL of 10 mM aqueous CuCl solution and 4 µL of 30% H_2O_2 to 947.5 µL of pH 7.40 bicarbonate solution (25 mM NaCl, 6.25 mM $NaHCO_3$). This mixture was pre-incubated for 10 min at room temperature; then, 41 µL of 3 mM salicylic acid aqueous solution was added, and the reaction was carried out for 2 min at room temperature. A volume of 200 µL of 4% phosphoric acid was added to stop the reaction. Immediately afterward, an aliquot of 100 µL of the solution was removed and transferred into a vial for HPLC-ED analysis.

3.3.2. HPLC-ED Conditions

The chromatographic system consisted of a model PU-1580 HPLC pump from Jasco (Tokyo, Japan) coupled to an ESA (Chelmsford, Massachusetts, MA, USA) coulometric Coulochem III detector, equipped with a high-sensitivity analytical cell with two porous graphite working electrodes placed in series in the same cell. The reference electrodes were α-hydrogen/palladium, and the support electrodes were made of 501 stainless steel. In the analytical cell, potential 1 was set to −200 mV and potential 2 was set to +450 mV, with a range of 200 nA and a +1.00 V output. The coulometric detector was also equipped with a conditioning cell set to a potential of +50 mV. The chromatograms were acquired in oxidation mode. Electrodes were cleaned at the end of each working week in order to prolong their use time and to obtain reproducible results. They were kept at +900 mV for 2 min to restore them, thoroughly oxidizing the components potentially fouling the electrodes. A Waters Cortecs C18 column (100 mm × 2.1 mm, 2.7 µm) equipped with a Waters Cortecs C18 VanGuard guard column (5 × 2.1 mm, 2.7 µm) was used as the stationary phase. The mobile phase was a mixture of triethylammonium phosphate buffer (50 mM; pH 2.00) and acetonitrile (88/12, *v*/*v*). HPLC analysis was carried out in isocratic mode at a flow rate of 1.0 mL/min, using an injection volume of 50 μL. The data integration system was the DataApex (Petrzilkova, Prague, Czech Republic) Chromatography Station software (version CSW32 1.4).

3.4. Methods Application to Fruit Juice Samples

3.4.1. Electrochemical Method

The 11 kinds of commonly consumed fruit juices were purchased from local markets in Bologna and were randomly selected based on their contents of vitamin C and the kind of fruit that is known to contain high concentrations of antioxidant compounds. In total, 50 mL aliquots of the juices were accurately drawn and centrifuged at 6000 rpm for 20 min.

Each supernatant was immediately analyzed after proper dilution with ultrapure water, typically using a dilution ratio of 1:1000. The values of AOC of the fruit juices were expressed, taking Vitamin C as a reference, to give rise to vitamin C equivalents (VCE). VCE indexes were calculated on the basis of the vitamin C calibration line. Analyses were carried out five times for each sample. The AOC values of the 11 fruit juices were also determined by means of three commonly used methods, i.e., ABTS, DPPH and ORAC, according to an already published procedure [5].

3.4.2. Chromatographic Method

To test the inhibition or stimulation of the •OH formation of the tested juices, each sample was diluted 1:100 (V/V) with ultrapure water; then, an aliquot of 7.5 μL of the diluted sample was added to the reaction mixture, replacing an equal volume of bicarbonate solution. After being subjected to Fenton's reaction in the presence of salicylic acid, the mixture was injected into the HPLC-ED system. The concentrations of the three analytes (catechol, 2,5-DHBA and 2,3-DHBA) were determined separately by interpolation on the respective calibration curves and then added together; the obtained sum was compared to the sum of the analyte concentrations obtained without any tested fruit juice. The result was expressed as the percentage of inhibition of •OH production in comparison to that of the control sample not containing the tested sample. Analyses were carried out five times for each sample.

4. Conclusions

The data coming from the analysis of 11 fruit juices by means of the two original analytical methods, previously developed and validated by our groups, for evaluating their antioxidant capacity are in good agreement, especially if both employ Fenton's reaction as the source of OH radicals. In such a case, even if the AOC is calculated and expressed in a different way by the electrochemical and chromatographic methods, three groups of samples are identified, which are characterized by a high, intermediate and low value of antioxidant capacity.

Therefore, we can state that the two analytical approaches can be considered alternatives to estimate the antioxidant capacity of real samples, since both measure the ability of antioxidants to block OH radical formation. Furthermore, they can be obviously exploited to investigate both the antioxidant and prooxidant capacity of significant bioactive compounds, natural extracts and foods in general.

Author Contributions: Conceptualization: D.T., L.M. and I.G.; Data curation: S.Z. and M.P.; Funding acquisition: D.T. and L.M.; Investigation: I.G., M.P. and R.M.; Methodology: M.P., S.Z. and I.G.; Supervision: D.T. and L.M.; Writing—original draft: D.T., L.M. and R.M. All authors have read and agreed to the published version of the manuscript.

Funding: D.T. acknowledges the support of Alma Mater Studiorum–University of Bologna. L.M. acknowledges the support of Erasmus + Programme of the European Union, Key Action 2: Strategic Partnerships, within the OEMONOM project (No. 2020-1-CZ01-KA203-078218).

Institutional Review Board Statement: Not applicable.

Informed Consent Statement: Not applicable.

Data Availability Statement: The data presented in this study are available on request from the corresponding author.

Conflicts of Interest: The authors declare no conflict of interest.

Sample Availability: Not applicable.

References

1. Neha, K.; Haider, M.R.; Pathak, A.; Yar, M.S. Medicinal prospects of antioxidants: A review. *Eur. J. Med. Chem.* **2019**, *178*, 687–704. [CrossRef] [PubMed]
2. Halliwell, B.; Murcia, M.A.; Chirico, S.; Aruoma, O.I. Free radicals and antioxidants in food and in vivo: What they do and how they work. *Crit. Rev. Food Sci. Nutr.* **1995**, *35*, 7–20. [CrossRef]
3. Munteanu, I.G.; Apetrei, C. A Review on Electrochemical Sensors and Biosensors Used in Assessing Antioxidant Activity. *Antioxidants* **2022**, *11*, 584. [CrossRef]
4. Haque, M.A.; Morozova, K.; Ferrentino, G.; Scampicchio, M. Electrochemical Methods to Evaluate the Antioxidant Activity and Capacity of Foods: A Review. *Electroanalysis* **2021**, *33*, 1419–1435. [CrossRef]
5. Gualandi, I.; Tonelli, D. A new electrochemical sensor for OH radicals detection. *Talanta* **2013**, *115*, 779–786. [CrossRef] [PubMed]
6. Gualandi, I.; Ferraro, L.; Matteucci, P.; Tonelli, D. Assessment of the Antioxidant Capacity of Standard Compounds and Fruit Juices by a Newly Developed Electrochemical Method: Comparative Study with Results from Other Analytical Methods. *Electroanalysis* **2015**, *27*, 1906–1914. [CrossRef]
7. Protti, M.; Gualandi, I.; Mandrioli, R.; Zappoli, S.; Tonelli, D.; Mercolini, L. Analytical profiling of selected antioxidants and total antioxidant capacity of goji (*Lycium* spp.) berries. *J. Pharm. Biomed. Anal.* **2017**, *143*, 252–260. [CrossRef] [PubMed]
8. Saracino, M.A.; Santarcangelo, L.; Raggi, M.A.; Mercolini, L. Microextraction by packed sorbent (MEPS) to analyze catecholamines in innovative biological samples. *J. Pharm. Biomed. Anal.* **2015**, *104*, 122–129. [CrossRef] [PubMed]
9. Saracino, M.A.; Mandrioli, R.; Mercolini, L.; Ferranti, A.; Zaimovic, A.; Leonardi, C.; Raggi, M.A. Determination of homovanillic acid (HVA) in human plasma by HPLC with coulometric detection and a new SPE procedure. *J. Pharm. Biomed. Anal.* **2006**, *42*, 107–112. [CrossRef] [PubMed]
10. Catapano, M.C.; Protti, M.; Fontana, T.; Mandrioli, R.; Mladěnka, P.; Mercolini, L. An Original HPLC Method with Coulometric Detection to Monitor Hydroxyl Radical Generation via Fenton Chemistry. *Molecules* **2019**, *24*, 3066. [CrossRef] [PubMed]

molecules

MDPI

Article

Antimony in Polyethylene Terephthalate-Bottled Beverages: The Migration Puzzle

Sergio Carneado, José Fermín López-Sánchez and Ángeles Sahuquillo *

Analytical Chemistry Section, Faculty of Chemistry, Universitat de Barcelona, Martí i Franquès 1-11, 08028 Barcelona, Spain; sergiocarneado@gmail.com (S.C.); fermin.lopez@ub.edu (J.F.L.-S.)
* Correspondence: angels.sahuquillo@ub.edu

Abstract: A novel strategy to assess the main variables that potentially affect the migration of antimony from PET bottles to beverages, including mineral waters and juices, is herein proposed. In a preliminary step, an LC-ICP-MS method previously used for water analysis was optimized to correct identify Sb species present in the studied matrices using HRMS. Subsequently, the influence of temperature and storage time up to 30 days on Sb migration from PET bottles into peach and pineapple juices of the same brand was studied. Storing PET bottled drinks at elevated temperatures (i.e., in a hot car or in summer) can cause antimony migration to exceed the limits allowed in the EU or USA. Because the behavior observed differed from the results reported for Sb migration in mineral waters, a second approach was proposed: three mineral water and two juice samples were kept in different PET containers and stored at an elevated temperature (up to 60 °C) to understand the role of the PET type and matrix simultaneously. This study demonstrated that both matrix characteristics and type of PET bottle greatly influence antimony leaching, highlighting the need to consider these variables together when conducting migration experiments. The obtained results can be helpful for developing future legislation concerning migration of pollutants from packing to food commodities.

Keywords: antimony migration; antimony speciation; juice; water; polyethylene terephthalate bottle; cross-migration experiments

Citation: Carneado, S.; López-Sánchez, J.F.; Sahuquillo, Á. Antimony in Polyethylene Terephthalate-Bottled Beverages: The Migration Puzzle. *Molecules* **2023**, *28*, 7166. https://doi.org/10.3390/molecules28207166

Academic Editor: Katarzyna Pawlak

Received: 14 September 2023
Revised: 16 October 2023
Accepted: 16 October 2023
Published: 19 October 2023

1. Introduction

The use of polyethylene terephthalate (PET) for food packaging is growing faster than that of any other plastic. Antimony trioxide (Sb_2O_3) is a common substance used in the manufacture of PET and is the most important catalyst used in the process. One of the proposed mechanisms for the incorporation of antimony into PET during the polycondensation process involves the conversion of Sb_2O_3 into Sb-glycolate in different forms [1]. However, X-ray absorption fine-structure spectroscopy of PET material showed that Sb(III) can be partially oxidized (up to 50%) to Sb(V), which may be more soluble than trivalent species [2]. Thus, according to the different proposals suggested by these authors, both Sb(III) and Sb(V) forms could be present in the PET surface and be in direct contact with packaged foodstuffs, even though polymer characteristics and PET packaging production can also play a role on antimony diffusivity [3].

The toxicity of antimony depends on its chemical form and oxidation state. Between the inorganic forms, Sb(III) is ten times more toxic than Sb(V) [4,5], whereas the organic antimony compounds are much less toxic. Thus, the United States Environmental Protection Agency (USEPA) [6] and the European Union (EU) [7] consider Sb a priority pollutant.

Total antimony contents from 100 to 400 mg kg^{-1} have been reported in PET containers irrespective of the commercial brand tested. In most cases, they are destined to the storage of mineral water, whereas in some works, no specification of the type of beverage or foodstuff to be stored is described [2,8–16]. Although these values are significantly high, the Sb concentrations in the corresponding bottled water samples are within the recommended values for drinking water. All the studies cited concluded that the presence of Sb in the bottled water was due to its migration from PET, as water obtained from a fresh source or bore or stored in plastic of other types, such as polypropylene (PP), low-density and high-density polyethylene (LDPE, HDPE), polystyrene (PS), or polycarbonate (PC), did not suffer from Sb contamination [8,9,17–19].

Antimony migration studies from PET into food products can provide useful information about the potential risk associated with its presence under different storage conditions. Several physicochemical factors, such as storage time, temperature, light irradiation, PET color, and matrix characteristics (e.g., pH and salt content), can influence Sb migration from PET containers to water samples [8–11,13,15,17–25]. However, there is a wide variability in the effects of these factors on Sb migration [3], and the predominant role of the PET type or the water matrix is not clear [3,26,27].

Much less information has been reported about the total antimony contents in other beverages bottled in PET, which have more complex matrices that include considerable amounts of salts or organic acids, such as soft drinks, red wine, and fruit juices. Although some soft drinks present higher Sb concentrations than water samples [13], in some cases, the final concentrations have been reported to be within the maximum European limit for drinking water [13,28,29], while antimony concentrations in some fruit juices exceed the limit [30]. Hansen and co-workers mentioned that Sb concentrations can be influenced by the carbohydrate content of the juice and the number of days beyond the expiry date.

Few studies identify, quantify, and establish the behavior and fate of Sb species migrating from PET into beverages or other liquid foods. The presence of different Sb(III) or Sb(V) coordination complexes has been confirmed in fruit juices, yogurt, vinegars, and spirits stored in PET, using liquid chromatography–inductively coupled plasma–mass spectrometry (LC-ICP-MS) and liquid chromatography–electrospray–mass spectrometry (LC-ES-MS). The chemical forms found were uncomplexed Sb(V) as $Sb(OH)_6^-$ or highly stable complexes with different coordination numbers between antimony and either citrate, adenosine and lactate [31,32], acetate [33], and pyruvic acid and hydroxy ethane sulfonic anions [34]. The few migration studies were performed on carbonated and non-carbonated juices bottled in PET and in spirits [34,35], where the storage effect at 50 or 60 °C was assessed for 15 days. The increase in antimony concentration observed in the carbonated samples was attributed to the increase in the migration of Sb(III), whereas the increase observed in spirits was attributed to a simultaneous release of antimony from PET and a conversion of the Sb species.

Welle and Franz [12] developed a mathematical model to predict migration of antimony into beverages with increasing temperature and time by using only the experimental data obtained from migration assays with food simulants and considering different variables including total Sb concentration in the PET, bottle volume, wall thickness, energy, and diffusion coefficients. They found that the calculated antimony concentrations were in good agreement with the data from the literature, pointing out the role of the PET container in migration.

According to the current state of this topic, the objective of this study was to investigate the factors affecting the migration of antimony from PET to beverages considering the plastic type, kinetic factors (time and temperature), and drink matrix (water and juices). For this purpose, a new methodological approach was tested: Firstly, the effect of storing temperature and time were assessed in pineapple and peach juices bottled in PET, extending a previous study on PET bottled water samples [15]. Storage temperatures from 4 to 60 °C and storage times up to 60 days were considered. Secondly, to understand the roles of both PET-type and matrix, a cross-migration experiment exchanging matrices and containers was carried out. Moreover, as there is little information on antimony speciation, the determination of total Sb and its species was performed in all batches of experiments, covering both identification of Sb species and quantification.

2. Results

2.1. Determination of Total Sb in Juices

The results for total Sb content in the seven juices (apple, grape, pear, pineapple, peach, orange, and tomato) are summarized in Table 1.

Table 1. Total antimony concentration (mean ± standard deviation) in juices determined by ICP-MS using as pre-treatment a sample centrifugation (n = 3, from different replicates of every sample).

Juice	Concentration (μg L^{-1})	RSD (%)
Apple	0.343 ± 0.002	0.7
Grape	0.937 ± 0.018	1.9
Pear	0.522 ± 0.008	1.6
Pineapple	0.338 ± 0.002	0.5
Peach	0.273 ± 0.009	3.2
Orange	0.364 ± 0.005	1.3
Tomato	0.166 ± 0.007	4.2

As can be seen, Sb concentration levels were low in all cases but quantifiable (higher than the limit of quantification 0.02 μg·L^{-1}).

2.2. Sb Speciation in Juices

With the optimized chromatographic method, the seven juice matrices were quantified. The concentrations of the uncomplexed Sb(V) were near the limit of detection (0.01 μg L^{-1}) in all samples. The concentrations of Sb(III) ranged between 0.12 and 0.42 μg L^{-1}, with a mean concentration of 0.29 μg L^{-1} and relative standard deviation (RSD) values from 0.7 to 10% (n = 3).

2.3. Migration Experiments

The effect of different storage temperatures on the migration of total Sb and Sb speciation with time was studied on peach and pineapple juice samples.

At the beginning of the test (day 0), all of the samples (three bottles for each juice flavor) were analyzed in triplicate, and all of them contained significant amounts of total Sb concentrations (0.36 ± 0.01 for peach juice and 0.50 ± 0.02 μg Sb L^{-1} for pineapple juice (n = 9)).

For the samples stored at 4 °C and 20 °C, no variations in total Sb concentration were observed over time up to 30 days.

Results of the experiments performed on PET stored samples at 40 °C and 60 °C are represented in Figure 1, which summarizes total Sb concentrations in juices over the month of storage.

As speciation analysis is concerned, all the samples contained Sb(III) as the major species at any temperature during the whole experiment. The concentration levels followed the same tendency as described for total Sb analysis. Small amounts of uncomplexed Sb(V) were also observed near the limit of detection, along with some unidentifiable signals of very low intensity. Storage at 4 °C and 20 °C did not result in a significant evolution of Sb concentrations. Sb(III) concentrations in the samples stored at 4 °C and 20 °C were 0.13 ± 0.02 and 0.27 ± 0.04 μg Sb L^{-1} for peach and pineapple, respectively. Storage at 40 °C and 60 °C showed a slight increase in Sb(III) concentration up to day 30.

Sb(III) average concentrations in the samples stored at 4 °C and 20 °C were 0.13 ± 0.02 and 0.27 ± 0.04 μg Sb L^{-1} for peach and pineapple, respectively, together with the concentrations in the samples stored at 40 °C and 60 °C. It can be observed that the concentrations of Sb(III) are lower than those obtained in the total Sb determination: The peach samples reached Sb concentrations of 0.17 ± 0.01 and 0.18 ± 0.02 μg Sb L^{-1} at 40 °C and

60 °C, respectively, whereas for the pineapple samples, these values are 0.53 ± 0.04 and 1.02 ± 0.09 μg Sb L^{-1}, respectively. RSD values were ≤10% in all cases. Figure 2 shows the chromatograms obtained from PET stored peach and pineapple juice at 60 °C at different times, along with a blank for comparison purposes.

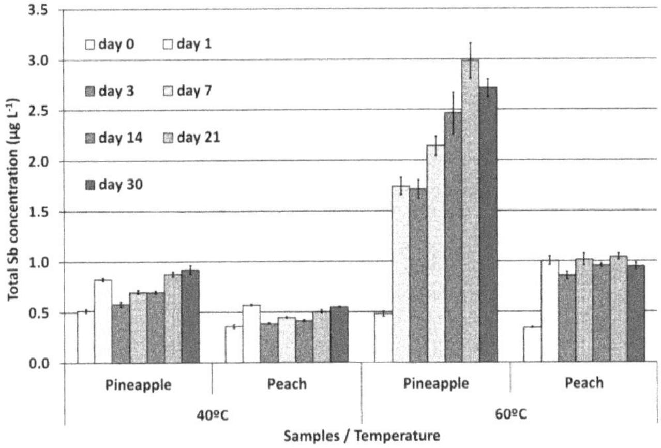

Figure 1. Total antimony concentration (μg·L^{-1}) in peach and pineapple juice PET-stored samples at 40 °C and 60 °C (mean (n = 9) ± standard deviation).

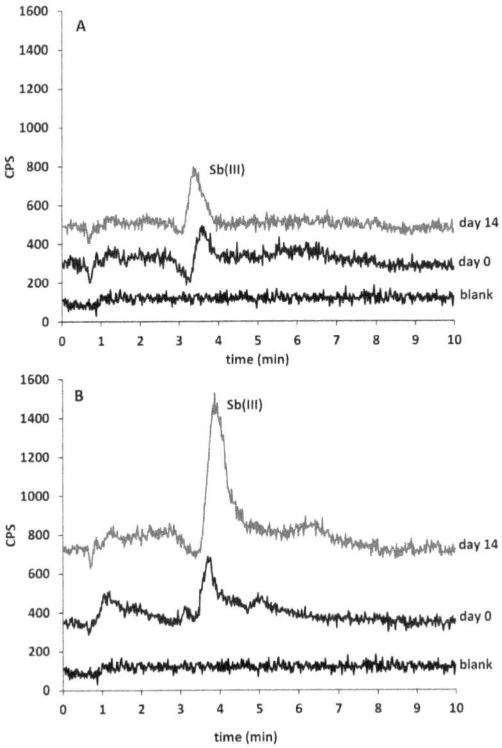

Figure 2. Chromatograms obtained by LC-ICP-MS from PET-stored peach juice (**A**) and PET-stored pineapple juice (**B**) at 60 °C at different days.

2.4. Cross-Migration Experiments

To study the extent to which Sb migration is governed by matrix and PET characteristics, a cross-migration experiment was carried out. The concentrations obtained at the beginning of the experiment (1.5 L bulk sample) (day 0) are summarized in Table 2. Total Sb concentrations in waters were slightly lower than those obtained in juices. Regarding speciation, the results obtained were the same as those obtained in the previous migration experiments: uncomplexed Sb(V) was the only species present in the water samples, whereas in juices, Sb(III) was the predominant species, and uncomplexed Sb(V) was present under the LOQ.

Table 2. Antimony concentrations obtained in water samples (WA, WB, and WC), and juice samples (pineapple, JD; peach, JE) at the beginning of the experiment (day 0). Results are expressed as mean Sb value ± standard deviation (n = 3), and the relative standard deviations are expressed in % in parentheses.

Sample	Total Sb (μg L^{-1})	Sb(V) (μg L^{-1})	Sb (III) (μg L^{-1})
WA	0.47 ± 0.03 (6.9%)	0.46 ± 0.02 (5.0%)	<LOD
WB	0.61 ± 0.04 (6.2%)	0.63 ± 0.06 (8.9%)	<LOD
WC	0.49 ± 0.05 (9.8%)	0.50 ± 0.05 (9.2%)	<LOD
JD	0.99 ± 0.11 (10.7%)	<LOQ	0.74 ± 0.04 (5.7%)
JE	1.45 ± 0.15 (10.1%)	<LOQ	1.09 ± 0.03 (3.0%)

Total Sb concentrations in the reference set of samples kept at 20 °C and after 30 days of crossed storage were not different from the initial contents, whereas for samples stored at 60 °C, some changes were observed already from day 7 of storage, as can be seen in Figure 3.

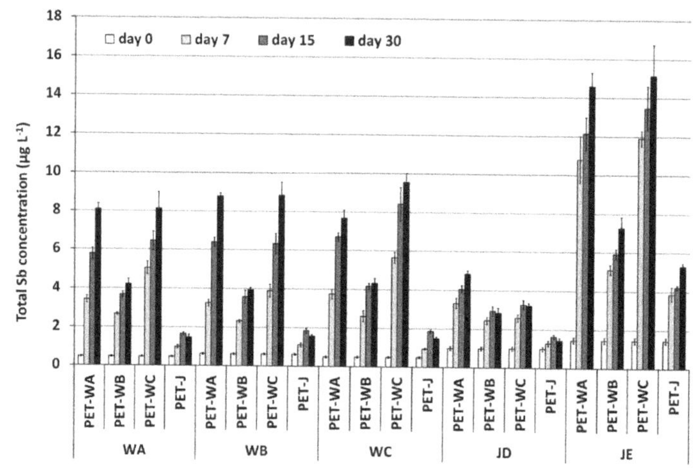

Figure 3. Total Sb concentrations (mean n = 3, standard deviation in bars) in mineral drinking waters and fruit juices samples used for the cross-migration experiment, stored at 60 °C for one month. In X-axis, the outer labels correspond to the beverage (water or juice), and the inner labels correspond to the PET bottle.

Regarding the speciation results, the antimony species concentrations obtained in the cross-migration experiment are summarized in Figure 4. On the X-axis, the outer labels correspond to the beverage (water or juice), and the inner labels correspond to the PET-bottle. For the groups of 4 bars grouped in each inner label, each column corresponds to the species concentrations measured at day 0, 7, 15, and 30, from left to right.

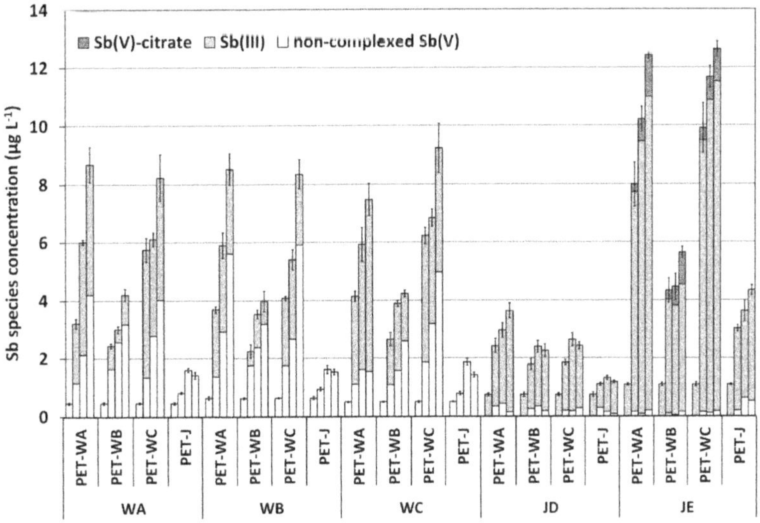

Figure 4. Sb species concentrations (mean $n = 3$, standard deviation in bars) in mineral drinking waters and fruit juices samples used for the cross-migration experiment, stored at 60 °C for one month.

Column recoveries of the sum of species calculated with respect to the total Sb concentration were between 80 and 100% for water samples and ranged between 70 and 85% for juices, which are acceptable for speciation in complex matrices.

3. Discussion

3.1. Determination of Total Sb in Juices

Independently of the juice flavor, the obtained concentrations of Sb were of the same order of magnitude in all samples and had good RSD values (≤5%). The maximum level of Sb established by the European Union in drinking water (5.0 µg L^{-1}) was not exceeded in any sample.

These results are comparable with those reported in the literature, with Sb concentrations ranging from 0.28 to 1.05 µg L^{-1} in orange and lemon juice contained in PET bottles 31 and 0.02 to 1.20 µg L^{-1} in apple, plum, and sour cherry juice [36]. Sb concentrations reported in 25 PET-bottled red fruit juices were lower than 0.5 µg L^{-1} in 10 of the samples analyzed, while 7 of them were above the maximum level established by EU for drinking water (5 µg L^{-1}) [30].

3.2. Sb Speciation in Juices: Optimisation of the Analytical Method

To start this study, an LC-ICP-MS method previously developed by the research group for water analysis was applied [15]. As it has been reported that different antimony species from those present in water may be present in juice because of the formation of complexes with the organic acids citric and malic [32,35], the separation process was assessed.

Firstly, to check the presence of antimony complexes in juice matrices, direct injection of pineapple and peach juices for high-resolution mass spectrometry (HRMS) was carried out. The mass spectra obtained showed high amounts of citric and malic acids but no Sb complexes. This is likely due to the very low concentration of antimony present. To observe the formation of Sb complexes, spiked samples were analyzed. In Sb(V)-spiked samples, the presence of the 1:1 citrate complex was detected, whereas in Sb(III)-spiked samples, the 1:2 citrate complex was observed. Moreover, the presence of malic complexes was not detected.

To identify the species occurring in the LC system, the analyses of individual Sb(V) and Sb(III) standard solutions of 0.25 µg L^{-1} diluted in the mobile phase (EDTA 10 mmol L^{-1})

and citric acid 40 mmol L^{-1} (simulating the concentration in juice matrix) were carried out. The results of analyzing the standards, depicted in Figure 5, showed that Sb species in the mobile-phase medium have different retention times from those diluted in citrate medium. This effect is due to the formation of different complexes between antimony and ethylenediaminetetraacetic acid (EDTA) or citrate, which have different structures and consequently retention capacities. When using EDTA standards, the retention times were as follows: 1 min for Sb(V), which corresponds to uncomplexed Sb(V), and slightly less than 4 min for Sb(III), which corresponds to an Sb(III)–EDTA complex [37]. When using citrate standards, the elution order is reversed: Retention times were slightly more than 4 min for Sb(III) and approximately 8 min for Sb(V). While the pentavalent species corresponds to an antimony–citrate complex [35], it is not clear whether Sb(III) corresponds to an antimony–citrate or antimony EDTA complex, as the retention times are similar.

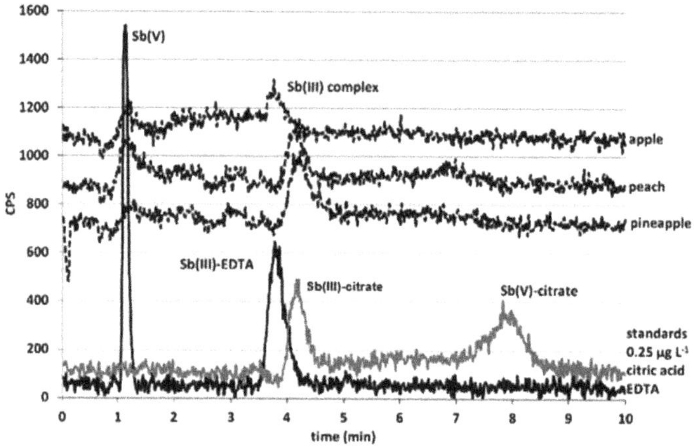

Figure 5. Chromatograms obtained from apple, pineapple, and juice samples together with antimony standards in mobile phase or citrate medium.

The analysis of the seven different-flavored PET-bottled commercial juices used in the previous section for the total Sb determination was carried out after centrifugation and filtration, as described above. Samples were analyzed with the conditions described in the Instrumentation section. As representative results, the upper part of Figure 5 depicts the chromatograms obtained for the pineapple, peach, and apple samples together with the chromatograms of the previously mentioned standards. Chromatographic peaks with the same retention times as Sb(III) standards were observed, and small amounts of uncomplexed Sb(V) were observed at 1 min.

To confirm the identity of the species observed in the juices corresponding to the complexes mentioned, spiking tests were performed. Samples were spiked with individual Sb(V) and Sb(III) standards diluted in both EDTA and citric acid. When Sb(V)-citrate was added, the 8 min peak corresponding to this standard appeared. Additionally, when the Sb(V) standard was added in EDTA media, a small peak corresponding to uncomplexed Sb(V) at a retention time of 1 min could be observed together with the peak of Sb(V)–citrate (approximately 70% of added Sb). This fact demonstrated that Sb(V) is complexed with citrate present in the juice.

Furthermore, the intensity of the main peak present in the samples (at approximately 4 min) increased when either Sb(III)–citrate or Sb(III)–EDTA was added. This behavior demonstrates that this peak corresponds to Sb(III) species, but it is still unclear whether the species observed in the chromatogram is an Sb(III)–citrate or Sb(III)–EDTA complex. However, it is suspected that this peak corresponds to an Sb(III)–EDTA complex, as it has been found in the literature that, although Sb(III)–citrate complexes are well known [31],

Sb(III) complexes convert to Sb(III)–EDTA complexes on the column when EDTA is part of the mobile phase [35].

Therefore, to check if the Sb(III) species observed in the standards prepared with citric acid and the juice samples corresponds to Sb(III)–EDTA or to Sb(III)–citrate complexes, two different approaches were carried out. The first approach was to inject decreasing volumes of 1 μg L^{-1} Sb(III) standards prepared in EDTA and citrate media and 1 μg L^{-1} Sb(III)-spiked pineapple and peach samples: 100, 50, 25, 10, and 5 μL. In the chromatogram profiles, the retention time of the EDTA media standard was the same at each volume injected. On the other hand, the retention time of the Sb(III) peak in the citrate media standard and the samples got closer to the retention time of the Sb(III)–EDTA standard as lower volumes were injected. From an injection volume of 10 μL, the retention time was the same as the Sb(III)–EDTA standard. As an example, Figure 6 depicts the chromatographic profiles of the standards prepared with citric acid and EDTA and the spiked samples when injecting volumes of 100 and 10 μL. Thus, it can be concluded that Sb(III) species eluted with an EDTA mobile phase from the standard prepared with citric acid and also from juice samples convert to Sb(III)–EDTA complexes. At higher working volumes and due to matrix effect, the Sb(III)–EDTA complex eluted at higher retention times.

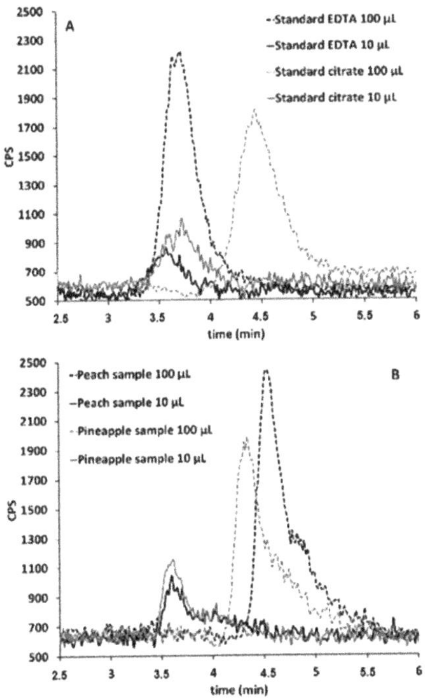

Figure 6. Chromatographic profiles obtained from 100 and 10 μL injections of 1 μg·L^{-1} standards of Sb(III) prepared in EDTA and citric media (**A**) and 1 μg·L^{-1} spiked peach and pineapple samples (**B**).

The conversion of Sb(III)–citrate species to Sb(III)–EDTA complexes was confirmed using high-resolution mass spectrometry (HRMS), by analyzing standards of Sb(III) in the presence of the stoichiometric concentration of EDTA and 10 times the stoichiometric concentration of citric acid for the correct formation of the complex [31], with the conditions explained in the Materials and Method section.

The Sb–citrate complex ($[Sb(citH_2)_2]^-$; $C_{12}H_{12}O_{14}Sb^-$), with an expected m/z value of 500.9260, was not detected, whereas the complex $[Sb(EDTA)]^-$ ($C_{10}H_{12}N_2O_8Sb^-$) was detected with a m/z value of 408.9634 with an error lower than 5 ppm. These results confirmed that trivalent forms of antimony are preferably complexed with EDTA instead of citrate in the presence of both substances in the conditions where separation is performed.

In conclusion, MS spectrometry analyses demonstrated that Sb(III) is present in samples as a 1:2 Sb(III)–citrate complex, which is converted into 1:1 Sb(III)–EDTA complex in the presence of the mobile phase during the analysis by LC-ICP-MS.

3.3. *Sb Speciation in Juices*

A dependent Student's *t*-test for paired samples, at the 95% confidence level, showed that the total Sb content was significantly higher than Sb(III) species content. This can be attributed to the small amounts of Sb species found in the measuring extracts, which were close to the limit of quantification, where the spread of expected results is high. Column recoveries ranged from 70 to 89%, which are acceptable for speciation in complex matrices, except for the grape sample (44%).

3.4. *Migration Experiments*

For total Sb, a two-way ANOVA at the 95% confidence level for the four temperatures and the seven periods of time tested indicated that both factors have an effect on the results.

When considering only the results obtained at 4 °C and 20 °C for all period of storage time assayed, a two-way ANOVA showed no effect of time and temperature on total Sb concentrations, and in no case did they exceed the maximum level established by the European Union for drinking water (5.0 $\mu g\ L^{-1}$). Samples stored at 40 °C showed a slight increase in total Sb concentration over time, reaching a concentration around $0.55 \pm 0.01\ \mu g\ L^{-1}$ for peach samples and $0.92 \pm 0.04\ \mu g\ L^{-1}$ for pineapple samples at 30 days of storage. The highest Sb migration was observed in the samples stored at 60 °C, which reached a concentration of $0.95 \pm 0.04\ \mu g\ L^{-1}$ in peach and $2.71 \pm 0.09\ \mu g\ L^{-1}$ in pineapple juice, after 21 days of storage.

The values obtained throughout the migration experiment for Sb(III), the only species present at concentrations higher than the LOQ, correspond approximately to 40–65% of the total antimony concentration for pineapple juice and to 20–45% of the total antimony concentration for peach samples. Hansen and Pergantis [35] obtained similar results, finding that the major species present in lemon and orange juices was Sb(III). According to their migration results at 50 °C, the antimony increase was mainly due to Sb(III), which is consistent with the results obtained in this study.

The lack of mass balance with respect to the total Sb content could be attributed to the minor signals that correspond not only to uncomplexed Sb(V) but also to unidentified antimony species. Apart from the increase of the main peak, Sb(III), with time, it can be visually observed that the baseline from the samples is not plain compared to that obtained from the blank. These slight increases in the background signal may be attributed to different Sb(V) complexes similar to those identified by Hansen and Pergantis [32] in spiked yogurts and juices, which cannot be separated correctly with the method conditions (strong anion exchange liquid chromatography). Moreover, specifically for peach juices, the Sb(III) signal is close to the limit of quantification, where the spread of expected results can be higher.

The only information regarding the effect of temperature on Sb migration in the literature is inconclusive, as only some juice samples stored at 50 °C for 14 days showed an increase in Sb levels [35]. In the present work, similar behavior was always observed at a fixed temperature.

The results indicate that temperature is the main factor that accelerates Sb migration into fruit juice. This effect can be attributed to the increased degradation of the PET with increasing temperature, which is well known, was described by Romao and co-workers [38], and was also observed in drinking water using X-ray absorption fine-structure (XAFS) techniques by Takahashi and co-workers [2]. Moreover, this degradation was also observed visually, as the sample bottles exhibited physical deformation. However, Sb migration also appeared to be dependent on the matrix, as it was more pronounced in pineapple than in peach juice, as can be observed in Figures 1 and 4. Comparing the migration data obtained in the present work about juices to those from a previous work about migration in PET-bottled mineral water [15], lower concentrations were observed for juice samples despite the expectation of higher concentration due to the high amount of salts and organic acids present in juices, which could have potentially increased Sb extractability from PET. Additional variables studied by Welle and Franz [12] can also affect the migration behavior. Total Sb concentration in PET, bottle volume and wall thickness, activation energy, and diffusion coefficients of Sb were used to propose a mathematical model for calculating migration with temperature and time. Our migration results with juices and mineral waters matched Welle's model, showing an analogous tendency even for the high temperatures not covered in their study.

In addition, Fan and co-workers [39] constructed kinetic curves for antimony release following the given equation:

$$C = C_{max} \cdot (1 - e^{-bt})$$

where "C" is the released antimony concentration at storage time "t" (days), "C_{max}" is the maximum concentration of released antimony (μg L^{-1}), and "b" is the kinetic constant (day^{-1}). Values of "C_{max}" and "b" can be estimated by fitting the data obtained in the equation [39]. Thus, the maximum concentration was released, and kinetic constants were calculated for the two fruit juice samples stored during 60 days at the four temperatures.

Kinetic curves at high temperatures matched the experimental values better, as evidenced by the higher R^2 values obtained. Considerable fewer differences in C_{max} and kinetic constants were obtained throughout the temperature range, which suggests that temperature is not the only variable that affects migration in this specific case.

Furthermore, C_{max} values (0.30–2.77 μg·L^{-1}) were of the same order of magnitude as the experimental maximum concentrations obtained (0.35–2.98 μg·L^{-1}). This demonstrates that all samples reached the maximum allowable migrated concentration. The Sb concentration in samples from day 30 onwards, if it had been possible to analyze them, should have been equal to or of the same order of magnitude as the results obtained in the kinetic curve showing that the maximum Sb concentration was reached. Regarding the kinetic constants (b), the values are of the same order in all cases (0.011–0.775) and are in a similar range as those obtained by Fan and co-workers [39] (0.19–0.65). This means that the migration process is similar at all temperatures. It is important to note that this migration behavior is different from that observed in mineral water samples [15]. This fact may be due to the differences between matrices and the PET containers in each case.

All these facts indicate that the antimony migration behavior is governed not only by the composition of the matrix but also by the characteristics of the PET container. This is the possible explanation for the results obtained in mineral waters and juices, which are packaged in different PET containers. Consequently, cross-migration experiments were designed.

3.5. Cross-Migration Experiments

For the results of total Sb in the reference samples stored at 20 °C after 30 days, a Student's *t*-test for means at 95% confidence level indicated no significant differences.

However, as expected, total antimony concentrations showed a significant increase throughout the days when stored at 60 °C. A two-way ANOVA test (95% confidence) confirmed the effect of storing time at this temperature. It can be observed that samples stored in bottles used for keeping mineral water (the first three groups of columns in Figure 3) exhibited more pronounced migration than those stored in bottles used for keeping juice (the last group of columns), as they reached higher Sb concentrations throughout the days. After 30 days of storage, mineral water samples stored in PET-WA and PET-WC bottles reached concentrations between 8 and 10 μg L^{-1}, while those stored in PET-WB bottles reached a concentration of approximately 4 μg L^{-1}, and those stored in PET-J bottles reached even a smaller concentration (around 2 μg L^{-1}).

Peach juice samples stored in PET-WA bottles reach a concentration of around 5 µg L^{-1}, while those stored in PET-WB and PET-WC bottles reach around 3 µg L^{-1}, and those stored in PET-J bottles reach less than 2 µg L^{-1}. Similarly, pineapple juice stored in PET-WA and PET-WC bottles reach approximately 15 µg L^{-1}, while those stored in PET-WB bottles reach around 7 µg L^{-1}, and those stored in PET-J bottle reach less than 5 µg L^{-1}.

Based on these results, it can be concluded that antimony migration depends on both the type of PET container and the sample matrix. Samples stored in PET-WA and PET-WC bottles exhibited the highest migration potential, followed by the PET-WB bottles, and then the PET-J bottles. Additionally, for each PET container, antimony migration was also dependent on the sample matrix. Antimony concentrations after 30 days of storage were lowest in peach juice samples, higher in mineral waters, and highest in pineapple juice samples.

Regarding antimony speciation (Figure 4), only non-complexed Sb(V) was present in water samples at the beginning of the experiment. After storage time, the concentration of non-complexed Sb(V) continued to increase, and Sb(III) species appeared in waters stored in all PET-bottle types. Sb(III) species were not observed in waters stored in the PET-J type.

For juice samples, non-complexed Sb(V) below the limit of quantification and Sb(III) were the only species present at the beginning of the experiment. After the storage at 60 °C, Sb(III) concentrations increased and remained the main species present in the samples. Non-complexed Sb(V) concentrations also increased, reaching values slightly above the limit of quantification. Additionally, in pineapple juice samples stored in PET water bottles, Sb(V)–citrate species appeared, reaching values between 1 and 1.5 µg L^{-1} at the end of the experiment.

4. Materials and Methods

4.1. Instrumentation

For total Sb determination by inductively coupled plasma–mass spectrometry (ICP-MS), an Agilent 7500 ce model with a reaction cell and Burgener Ari Mist HP nebulizer was used. More information about the ICP-MS system's operating conditions is summarized in Table 3.

Table 3. Operating conditions of the ICP-MS system.

ICP-MS Parameters	
RF power	1550 W
RF matching	1.66 V
Peristaltic pump speed	0.1 rps
Stabilization delay	30 s
Sampler and skimmer cones	Nickel
Nebulizer	BURGENER Ari Mist HP
Number of replicates	3
Spray chamber (type and temperature)	Scott-type and 15 °C
Carrier gas flow, Ar	0.75 L min^{-1}
Make up gas flow, Ar	0.33 L min^{-1}
Sampling depth	7.5 mm
Cell exit	−36 V
Quadrupole/Octopole bias difference	3 V
Mass to charge ratio	m/z 121 (^{121}Sb)

For Sb species determination, a quaternary pump Agilent 1200 (Waldbronn, Germany) equipped with an auto sampler and a Hamilton PRP-X100 (Reno, NV, USA) anion exchange column (125 × 4.1 mm, 10 µm particle size, USA) were used. This HPLC system was coupled to the ICP-MS, which operated under the following conditions: room temperature, 10 mM ethylenediaminetetraacetic acid (EDTA) at pH 4.0 with 0.5% MeOH as the mobile phase; 100 µL injection volume and 1.5 mL min^{-1} flow rate.

To characterize the structure of antimony complexes by high-resolution mass spectrometry (HRMS), a Thermo Fisher Scientific Q-Exactive Orbitrap (Madrid, Spain) equipped with a thermally assisted electrospray ionization source was used. The instrument was operated in negative mode with the following parameters: ESI voltage −2.5 kV; vaporizer and capillary temperatures 320 °C; sheath gas, auxiliary gas and sweep gas flow rates 40, 10, and 2 arbitrary units, respectively; and tube lens voltage 50 V. The mass spectrometers were operated in profile mode, with a scan range of m/z 100–1000 and a resolving power of 70,000 full width half maximum (FWHM) at m/z 200. An automatic gain control setting of 3×10^6 with a maximum injection time of 200 ms was used.

4.2. Reagents and Standards

Ultrapure water with conductivity of 5–15 M Ω cm^{-1} (Millipore, Bedford, MA, USA) was used for making up volume standards and reagents.

Standard solutions of 1000 mg L^{-1} Sb(III) and Sb(V) were prepared from $C_8H_4K_2O_{12}Sb_2$ xH_2O (Fluka, Neu-Ulm, Germany) and $KSb(OH)_6$ (Riedel de-Haën, Seelze, Germany), respectively. A certified standard solution of 1000 ± 4 mg L^{-1} antimony, prepared from 99.999% "purum" metallic Sb, dissolved and stabilized in high-purity acids (5% nitric acid (HNO_3) and 0.1% hydrofluoric acid (HF)), was used to standardize both standards. All standard solutions were kept refrigerated in high-density polyethylene bottles. The media of the daily working standard solutions were diluted acid and mobile phase for total and speciation Sb analysis, respectively.

A 10 mM EDTA (Panreac, Barcelona, Spain) solution adjusted to pH 4 with diluted ammonia (Panreac) was filtered daily through a 0.45 µm filter (Millipore type HA) for use as the mobile phase.

LC/MS grade methanol and water (Fluka, Madrid, Spain) were used to prepare the solutions for mass spectrometry analysis.

4.3. Selection of Samples

To characterize juices in preliminary experiments, seven commercial juices from the same brand were purchased from a local shop in order to cover the different flavors typically consumed in Spain and bottled in PET: apple, grape, pear, pineapple, peach, orange, and tomato. Juices were centrifuged and filtered through 0.45 µm, according to the procedure described by Welna and Szymczycha [36]. Then, 2 mL of each of the filtered juices was taken and diluted. Total and speciation analysis were performed by ICP-MS and LC-ICP-MS, respectively, as quantification techniques.

Two fruit juices, pineapple and peach, were selected for migration studies to study the influence of storage time and temperature on the Sb concentration in the juice. These juices are commonly consumed in Spain and were commercially available in suitable size batches.

For the cross-migration experiment, the same pineapple and peach juices were used together with three low-mineralized PET-bottled waters from different brands.

4.4. Migration and Cross-Migration Experiments

Three bottles of each flavor were stored at 4 °C, 20 °C, 40 °C, and 60 °C. Speciation and total antimony were analyzed in triplicate on seven occasions: before storage and after 1, 3, 7, 14, 21, and 30 days of storage. Longer storage was not carried out because, after one month of storage, juices at 20 °C appeared spoiled, while juices at 40 and 60 °C had darkened and become almost black, with a foul smell. Thus, because the juices were already decomposed after one month of storage, there was no point in continuing to analyze them for a longer time.

For the cross-migration experiment, three low-mineralized drinking waters (WA, WB, and WC) from different brands were selected based on their different PET container (shape, color, and thickness), along with peach and pineapple fruit juices (JD and JE, respectively) from the same brand. PET containers for juices had a different shape, were thinner than water bottles, and were colorless. For this experiment, five 0.33 L containers of each PET type were used (PET-WA, PET-WB, PET-WC, and PET-J). The contents of each container were discarded, and bottles were cleaned three times with double deionized water and dried at room temperature. Subsequently, each set of containers (e.g., PET-WA) was filled with all studied matrices (WA, WB, WC, JD, and JE), meaning that each matrix was placed inside all types of PET containers studied. For filling, a 1.5 L container of the studied matrices (waters and juices) was used. The content of the 1.5 L containers was also characterized in terms of total and speciation Sb content.

After filling, the bottles were stored at 60 °C for 30 days. Total Sb and speciation were analyzed in triplicate after 7, 15, and 30 days of storage. For reference, another set of bottles prepared as previously described was kept at 20 °C, and total Sb content was measured after 30 days of storage.

4.5. *Determination of Antimony by ICP-MS*

4.5.1. Total Analysis

A 2 mL aliquot of the sub-sample was diluted to 4 mL in HNO_3 1 mol L^{-1}. The samples were introduced into the ICP-MS system under the conditions described in the Instrumentation section. The samples were quantified using external calibration curves with total Sb standards from 0.1 to 7.5 µg L^{-1}. Each sample was analyzed in triplicate. For quality control purposes, rhodium was added to the nebulizer chamber to detect potential signal drifts during the working session. Additionally, the calibration curves standards were run before and after each sample series. Multi-elemental quality control standard solutions at two concentrations were measured at the end of the sequence to ensure stable instrument sensitivity.

4.5.2. Speciation Analysis

A 100 µL aliquot of sub-sample was directly injected into the LC system under the conditions described in the Instrumentation section. External calibration was used for quantification purposes using Sb(III) and Sb(V) solutions. After identifying the Sb species in the extracts by comparing their retention times to those of the standards, quantification was performed in the working range 0.05 to 2.5 µg L^{-1} for each species. The instrument stability during the working sessions was checked by measuring two Sb quality control standard solutions at the end of the sequence, and the calibrating standards were also run before and after each series of analysis. Data were acquired using ICP-MS Chemstation software (version C.01.00), and the peak areas were integrated using ICP-MS Chromatographic Data Analysis software (version C.01.00).

The same separation and measuring conditions proposed for the analysis of Sb in waters by LC-ICP-MS [15] can be also applied to the analysis of Sb species in the proposed juices, with a 10 min analysis time. The limits of detection (LOD) for the method are 0.01 and 0.03 µg·L^{-1} for Sb(V) and Sb(III), respectively, and the limits of quantification (LOQ) are 0.03 and 0.09 µg·L^{-1} for Sb(V) and Sb(III), respectively.

For quantification, standards prepared in the mobile-phase and citrate media are necessary because the presence of non-complexed Sb(V) and Sb(V)–citrate can be only quantified using the EDTA-media curve and citrate media curve, respectively. The trivalent species can be quantified with both Sb(III) calibration curves because the slopes obtained from the curves prepared with Sb(III)–EDTA and Sb(III)–citrate standards are not significantly different.

4.5.3. Structure Characterization by HRMS

A 2 µL aliquot of standard solution was filtered through a 0.22 µm filter and directly injected into the mass spectrometer operating under the conditions described in the Instrumentation section. Thermo Xcalibur Qual Browser software (version 3.1.) was used for instrument control and data acquisition.

4.6. Statistical Analysis

Students' *t*-test for comparing means, paired *t*-test for comparing series of data, and two-way ANOVA to ascertain the effect of factors affecting migration were carried out using Microsoft® Excel® 2016.

5. Conclusions

From the obtained results, it can be concluded that temperature is the main kinetic factor affecting antimony migration from PET to juice, as warmer sample storage leads to higher antimony migration from the bottle into the contents. Antimony (III) was the main species observed throughout the migration experiment.

Cross-migration experiments showed that antimony migration behavior can only be explained by considering both matrix composition (content) and PET bottle characteristics (container). The strategy used in this paper to study antimony migration proved to be a useful approach for assessing the possible role of the matrix and the PET type simultaneously.

The crucial role of the matrix composition in Sb migration from PET to food proven in this work should also be addressed in future updates to the present European legislation describing migration assays using food simulants such as water, acetic acid, and nitric acid. When complexing agents are present in the food matrices, as in the case of juice, simulants may underestimate the migrated amounts. Further work should ascertain the correlation between citric acid (and pH) and Sb concentration in the solvent in order to propose universal food simulants that improve food safety.

Author Contributions: Conceptualization, methodology, and supervision, Á.S. and J.F.L.-S.; investigation and formal analysis, data curation, visualization, and writing—original draft preparation, S.C.; writing—review and editing, S.C., Á.S. and J.F.L.-S.; resources and funding acquisition and project administration, J.F.L.-S. All authors have read and agreed to the published version of the manuscript.

Funding: This research was funded by the Grup de Recerca Consolidat (project number 2017SGR907), the Research Vice Rectorate of the University of Barcelona, and the Ministerio de Economia y Competitividad (Project CTQ2015-68685-R).

Institutional Review Board Statement: Not applicable.

Informed Consent Statement: Not applicable.

Data Availability Statement: Not applicable.

Acknowledgments: The authors thank Toni Padró (Centres Científics i Tecnològics of the University of Barcelona) for support with the ICP-MS analysis.

Conflicts of Interest: The authors declare no conflict of interest. The funders had no role in the design of the study; in the collection, analyses, or interpretation of data; in the writing of the manuscript; or in the decision to publish the results.

Sample Availability: Not applicable.

References

1. Duh, B. Effect of antimony catalyst on solid-state polycondensation of poly(ethylene terephthalate). *Polymers* **2002**, *43*, 3147–3154. [CrossRef]
2. Takahashi, Y.; Sakuma, K.; Itai, T.; Zheng, G.; Mitsunobu, S. Speciation of antimony in PET bottles produced in Japan and China by X-ray absorption fine structure spectroscopy. *Environ. Sci. Technol.* **2008**, *42*, 9045–9050. [CrossRef]
3. Filella, M. Antimony and PET bottles: Checking facts. *Chemosphere* **2020**, *261*, 127732. [CrossRef]
4. Filella, M.; Belzile, N.; Lett, M.C. Antimony in the environment: A review focused on natural waters III. Microbiota relevant interactions. *Earth-Sci. Rev.* **2007**, *80*, 195–217. [CrossRef]
5. Poon, R.; Chu, I.; Lacavalier, P.; Valli, V.; Foster, W.; Gupta, S.; Thomas, B. Effects of Antimony on Rats Following 90-Day Exposure via Drinking Water. *Food Chem. Toxicol.* **1998**, *36*, 21–35. [CrossRef] [PubMed]
6. EPA (Environmental Protection Agency). National Primary Drinking Water Regulations. Available online: https://www.epa.gov/ground-water-and-drinking-water/national-primary-drinking-water-regulations#Inorganic (accessed on 25 April 2023).

7. EU (Commission of the European Communities, Official Journal of the European Union). Directive 2003/40/EC. Available online: http://eur-lex.europa.eu/legal-content/EN/TXT/PDF/?uri=CELEX:32003L0040&from=EN (accessed on 25 April 2023).
8. Shotyk, W.; Krachler, M.; Chen, B. Contamination of Canadian and European bottled waters with antimony from PET containers. *J. Environ. Monit.* **2006**, *8*, 288–292. [CrossRef] [PubMed]
9. Shotyk, W.; Krachler, M. Contamination of bottled waters with antimony leaching from polyethylene terephthalate (PET) increases upon storage. *Environ Sci. Technol.* **2007**, *41*, 1560–1563. [CrossRef] [PubMed]
10. Westerhoff, P.; Prapaipong, P.; Shock, E.; Hillaireau, A. Antimony leaching from polyethylene terephthalate (PET) plastic used for bottled drinking water. *Water Res.* **2008**, *42*, 551–556. [CrossRef]
11. Keresztes, S.; Tatár, E.; Mihucz, V.G.; Virág, I.; Majdik, C.; Záray, G. Leaching of antimony from polyethylene terephthalate (PET) bottles into mineral water. *Sci. Total Environ.* **2009**, *407*, 4731–4735. [CrossRef]
12. Welle, F.; Franz, R. Migration of antimony from PET bottles into beverages: Determination of the activation energy of diffusion and migration modelling compared with literature data. *Food Addit. Contam.* **2011**, *28*, 115–126. [CrossRef]
13. Tukur, A.; Sharp, L.; Stern, B.; Tizaoui, C.; Benkreira, H. PET bottle use patterns and antimony migration into bottled water and soft drinks: The case of British and Nigerian bottles. *J. Environ. Monit.* **2012**, *14*, 1237–1247. [CrossRef] [PubMed]
14. Rungchang, S.; Numthuam, S.; Qiu, X.; Li, Y.; Satake, T. Diffusion coefficient of antimony leaching from polyethylene terephthalate bottles into beverages. *J. Food Eng.* **2013**, *115*, 322–329. [CrossRef]
15. Carneado, S.; Hernández-Nataren, E.; López-Sánchez, J.F.; Sahuquillo, A. Migration of antimony from polyethylene terephthalate used in mineral water bottles. *Food Chem.* **2015**, *166*, 544–550. [CrossRef] [PubMed]
16. Kishi, E.; Ozaki, A.; Ooshima, T.; Abe, Y.; Mutsuga, M.; Yamaguchi, Y.; Yamano, T. Determination of various constituent elements of polyethylene terephthalate bottles used for beverages in Japan. *Packag. Technol. Sci.* **2020**, *33*, 183–193. [CrossRef]
17. Ristić, M.; Popović, I.; Pocajt, V.; Antanasijević, D.; Perić-Grujić, A. Concentrations of selected trace elements in mineral and spring bottled waters on the Serbian market. *Food Addit. Contam. Part B Surveill.* **2011**, *4*, 6–14. [CrossRef]
18. Andra, S.S.; Makris, K.C.; Shine, J.P.; Lu, C. Co-leaching of brominated compounds and antimony from bottled water. *Environ. Int.* **2012**, *38*, 45–53. [CrossRef]
19. Magana-Maldonado, L.M.; Wrobel, K.; Espinoza Cruz, T.L.; Yanez Barrientos, E.; Corrales Escobosa, A.R.; Wrobel, K. Application of hydride generation—Microwave plasma—Atomic emission spectrometry and partial least squares regression for the determination of antimony directly in water and in PET after alkaline methanolysis. *Chemosphere* **2023**, *313*, 137316. [CrossRef]
20. Reimann, C.; Birke, M.; Filzmoser, P. Bottled drinking water: Water contamination from bottle materials (glass, hard PET, soft PET), the influence of colour and acidification. *Appl. Geochem.* **2010**, *25*, 1030–1046. [CrossRef]
21. Reimann, C.; Birke, M.; Filzmoser, P. Temperature-dependent leaching of chemical elements from mineral water bottle materials. *Appl. Geochem.* **2012**, *27*, 1492–1498. [CrossRef]
22. Bach, C.; Dauchy, D.; Severin, I.; Munoz, J.F.; Etienne, S.; Chagnon, M.C. Effect of temperature on the release of intentionally and non-intentionally added substances from polyethylene terephthalate (PET) bottles into water: Chemical analysis and potential toxicity. *Food Chem.* **2013**, *139*, 672–680. [CrossRef]
23. Aghaee, E.M.; Alimohammadi, M.; Nabizadeh, R.; Khaniki, G.J.; Naseri, S.; Mahvi, A.H.; Yaghmaeian, K.; Aslani, H.; Nazmara, S.; Mahmoudi, B.; et al. Effects of storage time and temperature on the antimony and some trace element release from polyethylene terephthalate (PET) into the bottled drinking water. *J. Environ. Sci. Health Eng.* **2014**, *12*, 133. [CrossRef]
24. Dogan, C.E.; Cebi, N. Investigation of antimony, cobalt, and acetaldehyde migration into the drinking water in Turkey. *Packag. Technol. Sci.* **2019**, *32*, 239–246. [CrossRef]
25. Chapa-Martínez, C.A.; Hinojosa-Reyes, L.; Hernández-Ramírez, A.; Ruiz-Ruiz, E.; Maya-Treviño, L.; Guzmán-Mar, J.L. An evaluation of the migration of antimony from polyethylene terephthalate (PET) plastic used for bottled drinking water. *Sci. Total Environ.* **2016**, *565*, 511–518. [CrossRef] [PubMed]
26. Rowell, C.; Kuiper, N.; Preud'Homme, H. Is container type the biggest predictor of trace element and BPA leaching from drinking water bottles? *Food Chem.* **2016**, *202*, 88–93. [CrossRef] [PubMed]
27. Gerassimidou, S.; Lanska, P.; Hahladakis, J.N.; Lovat, E.; Vanzetto, S.; Geueke, B.; Groh, K.J.; Muncke, J.; Maffini, M.; Martin, O.V.; et al. Unpacking the complexity of the PET drink bottles value chain: A chemicals perspective. *J. Hazard. Mater.* **2022**, *430*, 128410. [CrossRef] [PubMed]
28. De Jesus, A.; Dessuy, M.B.; Huber, C.S.; Zmozinski, A.V.; Duarte, A.T.; Vale, M.G.R.; Andrade, J.B. Determination of antimony in PET containers by direct analysis of solid samples using graphite furnace atomic absorption spectrometry and leaching studies. *Microchem. J.* **2016**, *124*, 222–227. [CrossRef]
29. Shakerian, F.; Dadfarnia, S.; Haji Shabani, A.M.; Nili Ahmad Abadi, M. Synthesis and characterisation of nano-pore antimony imprinted polymer and its use in the extraction and determination of antimony in water and fruit juice sample. *Food Chem.* **2014**, *145*, 571–577. [CrossRef]
30. Hansen, C.; Tsirigotaki, A.; Bak, S.A.; Pergantis, S.A.; Stürup, S.; Gammelgaard, B.; Hansen, H.R. Elevated antimony concentrations in commercial juices. *J. Environ. Monit.* **2010**, *12*, 822–824. [CrossRef]
31. Hansen, H.R.; Pergantis, S.A. Investigating the formation of an Sb(V)–citrate complex by HPLC-ICP-MS and HPLC-ES-MS(/MS). *J. Anal. At. Spectrom.* **2006**, *21*, 1240–1248. [CrossRef]
32. Hansen, H.R.; Pergantis, S.A. Identification of Sb(V) complexes in biological and food matrixes and their stibine formation efficiency during hydride generation with ICPMS detection. *Anal. Chem.* **2007**, *79*, 5304–5311. [CrossRef]

33. Sánchez-Martínez, M.; Pérez-Corona, T.; Cámara, C.; Madrid, Y. Migration of antimony from PET containers into regulated EU food simulants. *Food Chem.* **2013**, *141*, 816–822. [CrossRef] [PubMed]
34. Carneado, S.; López-Sánchez, J.F.; Sahuquillo, A.; Klontzas, E.; Froudakis, G.E.; Pergantis, S.A. Antimony speciation in spirits stored in PET bottles: Identification of a novel antimony complex. *J. Anal. At. Spectrom.* **2017**, *32*, 1109–1118. [CrossRef]
35. Hansen, H.R.; Pergantis, S.A. Detection of antimony species in citrus juices and drinking water stored in PET containers. *J. Anal. At. Spectrom.* **2006**, *21*, 731–733. [CrossRef]
36. Welna, M.; Szymczycha-Madeja, A. Effect of sample preparation procedure for the determination of As, Sb and Se in fruit juices by HG-ICP-OES. *Food Chem.* **2014**, *159*, 414–419. [CrossRef] [PubMed]
37. Filella, M.; May, P.M. Critical appraisal of available thermodynamic data for the complexation of antimony(III) and antimony(V) by low molecular mass organic ligands. *J. Environ. Monit.* **2005**, *7*, 1226–1237. [CrossRef]
38. Romao, W.; Franco, M.F.; Corilo, Y.E.; Eberlin, M.N.; Spinacé, M.A.S.; De Paoli, M. Poly(ethylene terephthalate) thermo-mechanical and thermo-oxidative degradation mechanisms. *Polym. Degrad. Stab.* **2009**, *94*, 1849–1859. [CrossRef]
39. Fan, Y.Y.; Zheng, J.L.; Ren, J.H.; Luo, J.; Cui, X.Y.; Ma, L.Q. Effects of storage temperature and duration on release of antimony and bisphenol A from polyethylene terephthalate drinking water bottles of China. *Environ. Pollut.* **2014**, *192*, 113–120. [CrossRef]

molecules

MDPI

Article

Applicability and Limitations of a Capillary-LC Column-Switching System Using Hybrid Graphene-Based Stationary Phases

João Victor Basolli Borsatto [1,2], Edvaldo Vasconcelos Soares Maciel [1,3], Alejandro Cifuentes [2,*] and Fernando Mauro Lanças [1,*]

[1] Laboratory of Chromatography, Institute of Chemistry at Sao Carlos, University of Sao Paulo, P.O. Box 780, Sao Carlos 13566590, Brazil; jvictorborsatto@usp.br (J.V.B.B.); daltoniqsc@gmail.com (E.V.S.M.)
[2] Laboratory of Foodomics, Institute of Food Science Research (CIAL, CSIC), Nicolás Cabrera 9, 28049 Madrid, Spain
[3] Clemens Schöpf Institute, Department of Chemistry, Technical University of Darmstadt, 64287 Darmstadt, Germany
[*] Correspondence: a.cifuentes@csic.es (A.C.); flancas@iqsc.usp.br (F.M.L.)

Abstract: Graphene oxide sheets fixed over silica particles (SiGO) and their modification functionalized with C18 and endcapped (SiGO-C18ec) have been reported as sorbents for extraction and analytical columns in LC. In this study, a SiGO column was selected as the extraction column and a SiGO-C18ec as the analytical column to study the applicability and limitations of a column-switching system composed exclusively of columns packed with graphene-based sorbents. Pyriproxyfen and abamectin B1a were selected as the analytes, and orange-flavored carbonated soft drinks as the matrix. The proposed system could be successfully applied to the pyriproxyfen analysis in a concentration range between 0.5 to 25 μg/mL presenting a linearity of R^2 = 0.9931 and an intra-day and inter-day accuracy of 82.2–111.4% (RSD < 13.3%) and 95.5–99.8% (RSD < 12.7%), respectively. Furthermore, the matrix composition affected the area observed for the pyriproxyfen: the higher the concentration of orange juice in the soft drink, the higher the pyriproxyfen the signal observed. Additionally, the SiGO extraction column presented a life use of 120 injections for this matrix. In contrast, the proposed system could not apply to the analysis of abamectin B1a, and the SiGO-C18ec analytical column presented significant tailing compared to a similar approach with a C18 analytical column.

Keywords: capillary liquid chromatography; graphene-based materials; packed column; column switching; direct injection; SiGO; SiGO-C18ec

Citation: Borsatto, J.V.B.; Maciel, E.V.S.; Cifuentes, A.; Lanças, F.M. Applicability and Limitations of a Capillary-LC Column-Switching System Using Hybrid Graphene-Based Stationary Phases. *Molecules* **2023**, *28*, 4999. https://doi.org/10.3390/molecules28134999

Academic Editors: Gavino Sanna and Petr Bednar

Received: 12 May 2023
Revised: 15 June 2023
Accepted: 19 June 2023
Published: 26 June 2023

1. Introduction

Graphene and graphene-based materials are used in diverse analytical chemistry applications [1–3]. These materials have also been receiving increasing attention in sample preparation and liquid chromatography (LC) in the last decade. Graphene-based phases can be present in different ways. Fe_3O_4 magnetic particles containing graphene [4] or graphene oxide [5], graphene-based aerogels [6,7] and hydrogels [8,9], and graphene-based material fixed on silica particles [10–12] are examples of how these materials can be produced. Other possibilities of application for graphene-based materials are summarized in recent reviews [13–15].

Applications in separation science are as diverse as the possibilities of graphene-based material compositions [16–18]. Techniques such as stir bar sorptive extraction [16,19], solid-phase extraction [10,20], fiber solid-phase microextraction [20,21], in-tube solid-phase microextraction [12,22], and others [1,23,24] are examples of graphene-based material applications in sample preparation techniques. A particular focus can be given to column-switching methods coupled with LC or miniaturized LC because they minimize the steps of sample preparation procedures [25,26]. Sample preparation errors are one of the three most predominant sources of error in analytical chemistry [27]. Among the graphene-based phases applied to column-switching, graphene oxide sheets fixed to amino-silica particles (SiGO) and graphene oxide sheets fixed to amino-silica particles functionalized with C18 and endcapped (SiGO-C18ec) can be highlighted. These phases have been successfully applied in environmental, biological, and food analysis [28,29]. For example, extraction columns (also known as pre-concentration columns) packed with SiGO-C18ec particles have been applied with success in studies of pesticides in sugarcane spirits [30], mycotoxin separation in beverages [31], and xanthines in coffee beverages [32]. Examples of their application in a column-switching setup containing SiGO particles in the extraction column are the determination of β-lactams from environmental water samples [33], pesticides in sugarcane spirits [30], and resolution of antidepressant and antiepileptic drugs in urine [34].

Graphene-based materials have also been explored as packing material for analytical columns in LC [28]. Graphene sheets fixed to silica particles [35,36], graphene oxide sheets fixed to silica particles functionalized with C18 [37,38], graphene sheets fixed to silica particles modified with gold nanoparticles [39], and cellulose-coated reduced oxide sheets fixed to silica particles [40] are examples of some materials that have been applied as stationary phases in liquid chromatography. Columns packed with the SiGO-C18ec phase have been successfully used in quantitative analyses of hormones in urine samples [41]. A recent study compared a SiGO-C18ec with a commercial C18 column, both packed capillary LC columns operating in reverse-phase LC [42]. It was observed that SiGO-C18ec phases present different selectivity from C18 phases. Once graphene-based columns are applicable as both extraction and analytical columns in column-switching mode, it is possible to presume that a column-switching system composed of only graphene-based columns could be viable. This work evaluates the applicability and limitations of a column-switching system consisting exclusively of columns packed with graphene-based stationary phases. To the best of our knowledge, no work has described the use of a column-switching system composed exclusively of graphene-based packed columns in any application, and this a vacancy to be investigated. For this study, a SiGO column was selected as the extraction column and a SiGO-C18ec as the analytical column. In addition, pyriproxyfen and abamectin B1a were selected as analytes, and orange-flavored carbonated soft drinks as a matrix. Orange soft drink, are an interesting matrix to evaluate this proposed column-switching system. Though this matrix is complex and presents solid material in suspension, it can also be directly injected in the column-switching system without pretreatment. Pyriproxyfen and abamectin B1a are insecticides employed to control several pest species in diverse cultures. They have been selected as model analytes for this study because quantitative analysis has already been reported in the literature for both compounds in orange and orange by-product samples [43,44]. The leading figures of merit were evaluated, including linearity, accuracy, and precision of the calibration curve, the matrix effect, and the column life-use parameters. A comparison of the peak shape with a C18 analytical column is presented.

2. Results and Discussion

2.1. Selection of the Analytes

Pyriproxyfen and abamectin B1a are pesticides from two distinct chemical classes, which present different structures (Figure 1).

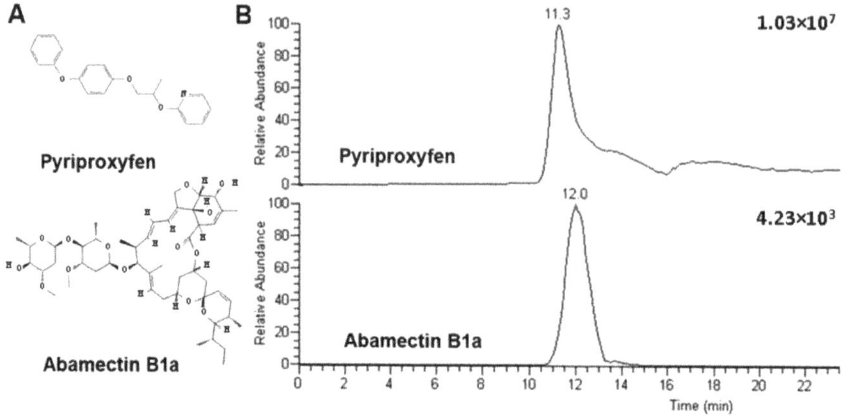

Figure 1. Structure of the analytes evaluated (**A**) and chromatograms obtained in the column-switching method described (**B**). Separation conditions: loading time of 3 min at 50 µL/min, linear gradient from 50% to 100% B in 15 min followed by an additional 5 min of 100% of B; 100 µL of the sample at a concentration of 10 µg/mL of each analyte was injected.

Pyriproxyfen mainly comprises aromatic rings and abamectin B1a by aliphatic chains (Figure 1A). Despite these structural differences, pyriproxyfen and abamectin B1a presented closed retention times in the column-switching system composed of the columns packed with SiGO and SiGO-C18ec (Figure 1B). The similarity in the retention time indicated that both analytes were similarly retained in the proposed column-switching system. This characteristic was important to evaluate whether the proposed column-switching system, composed exclusively of columns packed with graphene-based phases, is multipurpose or not. If the system demonstrated successful quantitative analysis for only one of the analytes, it would suggest its limitations as a multipurpose system. Conversely, if it provided suitable quantitative analysis for both compounds, further investigation would be necessary to confirm the multipurpose capabilities of the system. Because of the characteristics mentioned above, these compounds were selected for this study. In addition, the gradient method was optimized to ensure the elution of the monitored compounds and to elute non-monitored compounds that could stay retained in the columns and reduce the life use of the devices (Section 2.6 details the life-use aspects of these columns). Figure 1B depicts two other pieces of evidence: (i) the pyriproxyfen MRM chromatogram presents a more intense signal than the abamectin 1Ba MRM chromatogram; and (ii) the pyriproxyfen MRM chromatogram presents a more pronounced tail, which is almost not present in the abamectin B1a chromatogram.

2.2. Loading Method

The first variable evaluated for this study was the loading flow rate. The loading flow rate directly affects the extraction of the analytes in the extraction column. Figure 2 shows the area and the relative standard deviation (% RSD) obtained for both evaluated compounds at 25, 50, and 100 µL/min loading flow rates.

The loading flow rate did not affect the area obtained for the pyriproxyfen (Figure 2A), but a reduction in the % RSD was observed with the decrease in the loading flow rate (Figure 2B). For the abamectin B1a, the 50 µL/min loading flow rate resulted in a higher area (Figure 2C) and lower relative % RSD (Figure 2D) than the other explored. Therefore, the 50 µL/min loading flow rate was selected to be used this study because it allowed for better extraction of both analytes from the matrix.

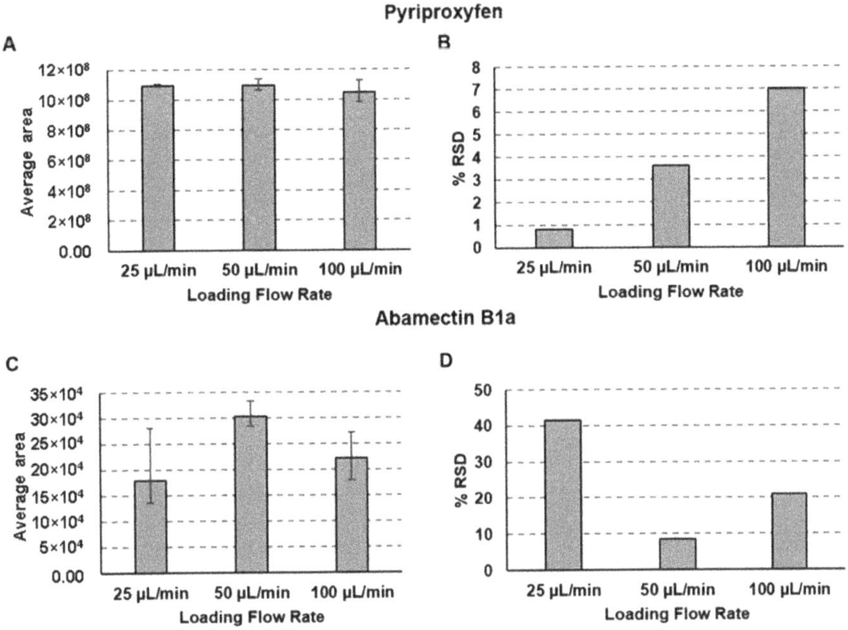

Figure 2. The area obtained with loading flow rates of 25, 50, and 100 µL/min (**A**) and % RSD for each loading flow rate (**B**) for pyriproxyfen and the area obtained with loading flow rates of 25, 50, and 100 µL/min (**C**) and % RSD for each loading flow rate (**D**) for abamectin B1a. Separation conditions: loading time of 3 min at 50 µL/min, linear gradient from 50% to 100% B in 15 min followed by an additional 5 min of 100% of B; 100 µL of the sample at a concentration of 10 µg/mL of each analyte was injected. The analyses were performed in triplicate for each flow rate, and the error bars represent the maximal and minimal values observed experimentally.

2.3. *Linearity of the Calibration Curve*

Calibration curves are the core of quantitative analysis [45]. The experimental data in Figure 3 shows that the column-switching method presented an excellent linearity for the study of pyriproxyfen, with $R^2 > 0.99$ (Figure 3A).

The % RSD for the pyriproxyfen analysis was lower than 12% for all the concentrations (Figure 3B). Both observations suggest that the experimental setup composed of a first column containing SiGO (extraction column) and a second column of SiGO-Ç18ec (analytical column) is suitable for analyzing pyriproxyfen in beverages consisting of orange juice. On the other hand, for abamectin B1a, the calibration curve did not present good linearity, with $R^2 < 0.47$ (Figure 3C), and gave high values for % RSD (Figure 3D). Based on that observation, it is possible to point out one of the limitations of the explored system: SiGO columns are not multipurpose, at least not at this point of development, and their application could still be limited. These limitations are already expected once graphene-derived columns are under early development. Usually, each kind of reverse-phase stationary phase has a specific range of operations. In some cases, it can be broad (C18, as an example) but limited in others (chiral phases, as an example). Although it demands much more investigation before a conclusion can be drawn, it seems that the graphene-based stationary phase can become one option for specific applications instead of a multipurpose phase.

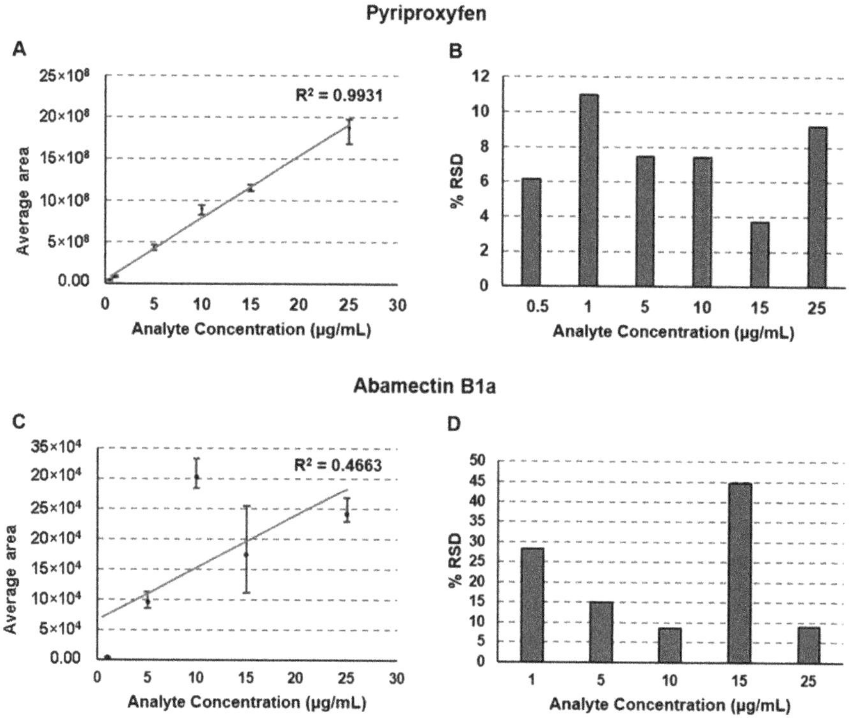

Figure 3. The calibration curve (**A**) and % RSD for each point of the calibration curve (**B**) for pyriproxyfen and the calibration curve (**C**) and % RSD for each point of the calibration curve (**D**) for abamectin B1a. Separation conditions: loading time of 3 min at 50 µL/min, linear gradient from 50% to 100% B in 15 min followed by an additional 5 min of 100% of B; 100 µL of the sample was injected. The analyses were performed in triplicate for each concentration and the error bars represent the maximal and minimal values observed experimentally.

2.4. Intra-Day and Inter-Day Accuracy and Precision

Accuracy and precision are other critical factors in quantitative analysis. For this work, the acceptance criteria were accuracy between 70–110% and precision with % RSD <20%; these values were based on previous work using graphene-based extraction or analytical columns [33,41]. Table 1 presents the accuracy and precision data obtained. For the pyriproxyfen analysis, the intra-day % RSD was lower than 20% for all the points and, in most determinations, lower than 10%.

The inter-day % RSD for this compound was 12.7%, 3.9%, and 2.9% for 2, 8, and 20 µg/mL concentrations, respectively. The intra-day accuracy for the analysis of the pyriproxyfen was between 82.2% and 111.4%, and the inter-day accuracy was between 95.5% and 99.8%. For the analysis of abamectin B1a, the proposed column-switching system was demonstrated not to be reliable. For this compound, obtaining intra-day and inter-day % RSD higher than 20% and accuracy in a wide range between 5.8% and 207% was typical. The observations stated in the last section are confirmed here. Graphene-based stationary phases are promising phases that can be used in a column-switching system, but their applications might not be multipurpose and present limitations.

Table 1. Intra-day and inter-day accuracy and precision (% RSD) obtained using the described column-switching system for both analytes. Separation conditions: loading time of 3 min at 50 µL/min, linear gradient from 50% to 100% B in 15 min followed by an additional 5 min of 100% of B; 100 µL of sample was injected.

Compound		Pyriproxyfen			Abamectin B1a		
Concentration (µg/mL)		2	8	20	2	8	20
Expected Area *		190,000,000	610,000,000	1,450,000,000	84,362	136,104	239,587
Day 1	Area	176,290,988	612,883,132	1,438,632,466	4899	120,972	274,393
	Intra-Day RSD %	5.8	7.8	3.1	172.8	22.6	9
	% Accuracy **	92.8	100.5	99.2	5.8	88.9	114.5
Day 2	Area	156,108,411	621,257,647	1,501,904,177	7836	99,444	378,336
	Intra-Day RSD %	8.9	13.3	3.3	96.1	31.3	3
	% Accuracy **	82.2	101.8	103.6	9.3	73.1	157.9
Day 3	Area	211,749,192	567,532,037	1,399,813,509	12,072	282,730	372,730
	Intra-Day RSD %	3.2	5.3	5.4	344.6	18.9	17
	% Accuracy **	111.4	93.0	96.5	14.3	207.7	155.6
Average Inter-Day Area		181,382,863	600,557,605	1,446,783,384	8269	167,715	341,820
Inter-Day RSD % ***		12.7	3.9	2.9	35.6	48.8	14
Inter-Day % Accuracy **		95.5	98.5	99.8	9.8	123.2	142.7

* Calculated based on the calibration curve. ** Accuracy is calculated based on the expected area from the calibration curve. *** Calculated based on the intra-day averages.

2.5. Matrix Effects

The matrix composition might affect the analytical signal (peak area, as an example) observed, interfering with the analysis. This is called the "matrix effect". Knowledge about the matrix effect is essential in most analysts' chromatography column-switching approach to extraction, especially when a sample pretreatment is not included in the analytical workflow before the injection step. To conduct this experiment, two commercial orange-flavored soft drinks with different proportions of orange juice in their compositions and spiked with pyriproxyfen in water were explored. Only pyriproxyfen was used to study the effect of the matrix. In the concentration of 2 µg/mL, the impact of the matrix was minimal. However, the analytes might have been better extracted in orange juice samples than pure water (Figure 4A).

The same trend was observed for the concentration of 8 µg/mL: the extraction in both samples containing orange juice was similar, and both were better than the aqueous sample (Figure 4B). For the 20 µg/mL solution, it is noted that the higher the percentage of orange juice in the matrix composition, the higher the area of the pyriproxyfen obtained (Figure 4C). These observations allow us to conclude that the matrix affects the extraction capability of the SiGO column, and this effect is more perceptible as the concentration of the analytes increases. These observations are even more visible in Figure 4D, which summarizes the average area obtained for the samples containing 2, 8, and 20 µg/mL of pyriproxyfen.

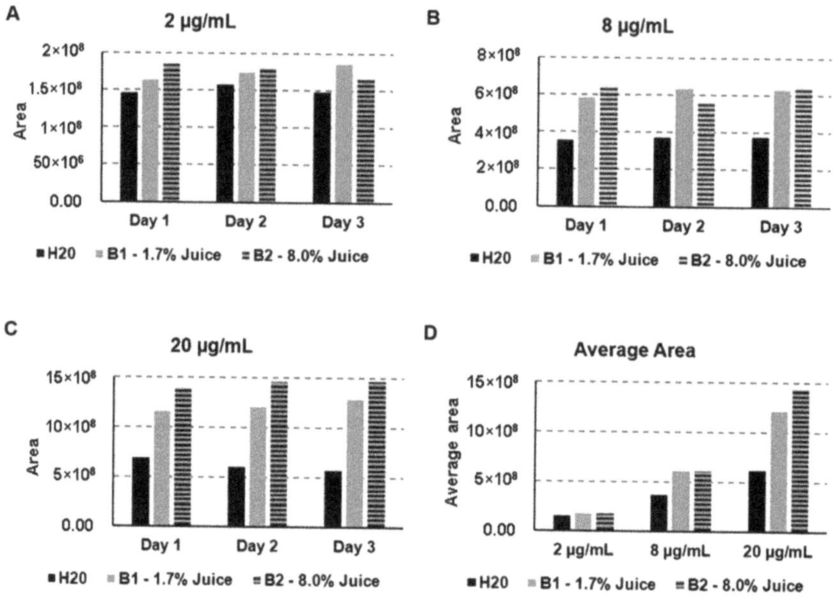

Figure 4. Matrix effect in the obtained area of pyriproxyfen for the 2 µg/mL (**A**), 8 µg/mL (**B**), 20 µg/mL (**C**), and a comparison of the average for the three concentrations (**D**). Separation conditions: loading time of 3 min at 50 µL/min, linear gradient from 50% to 100% B in 15 min followed by an additional 5 min of 100% of B; 100 µL of analytes were injected.

2.6. System Life Use

As yet, the life use of graphene-based columns is uncertain. In this study, we hope to bring some light to this aspect. The analyses were performed by injecting the sample without any preparation (despite adding 25% acetonitrile to keep the analytes soluble). After its hundredth analysis, the SiGO column (extraction column) presented for the first time a carry-over effect (Figure 5). Additionally, the column showed a blockage after about the 120th analysis. The column was washed in the back flush mode, removing the blockage, but the carry-over effect remained.

Additionally, after the washing, the column lost its repeatability compared to the previous separations. Therefore, the column was not considered helpful after this for this application. In contrast, although the SiGO-C18ec column (analytical) has an unknown life-use time, it has been used in uncountable separations (at least more than 300) and remains usable. It is essential to highlight that the SiGO-C18ec column used in this study has been used only as the analytical column, which may favor its durability.

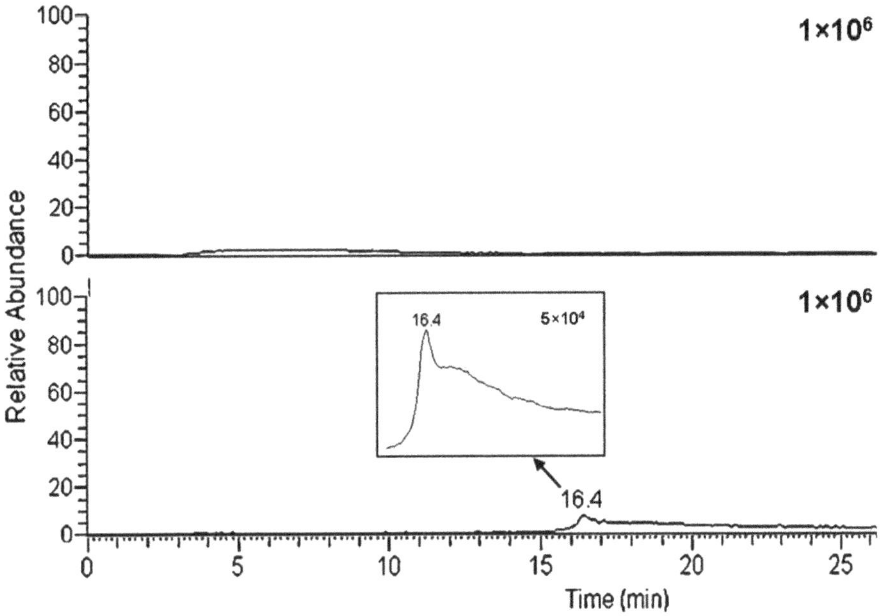

Figure 5. Blanks were obtained after 50 and 100 analyses performed in the same column. Separation conditions: loading time of 3 min at 50 µL/min, linear gradient from 50% to 100% B in 15 min followed by an additional 5 min of 100% B. No sample was injected.

2.7. Peak Shape Comparison with C18 Analytical Column

Another variable explored was the effect of the analytical column in the chromatograms obtained by the described column-switching method. The SiGO-C18ec (0.3 × 100 mm, 10 µm FPPs) column was compared to a C18 column (50 × 2.6 mm, 1.9 µm SPPs) using the same chromatographic method (Figure 6). For the abamectin B1a, it is clear that the C18 column resulted in thinner peaks than the SiGO-C18ec column for all the concentrations explored. No tail was observed for this compound in either column. However, a similar tailing was observed in both columns for the pyriproxyfen analysis, indicating that it may be caused in the SiGO extraction column. Although both columns presented tails for the pyriproxyfen peak, the tailing was higher (proportionally to the peak height) on the SiGO-C18ec column. In some separations using the SiGO-C18ec column, especially at concentrations $\geq$ 15 µg/mL, additional analysis was necessary to understand better and define when the tail ends and the baseline starts. One relevant highlight that must be taken into account here is the difference between the particle diameter and its porosity between the columns compared. The SiGO-C18ec is packed with 10 µm FPPs, while the C18 column is packed with 1.9 µm SPPs. Usually, reducing the particle diameter results in better performance, and reducing the particle core porosity does the same [46,47]. Additionally, as already mentioned, the SiGO-C18ec columns are in an early, underdevelopment stage, and their comparison to the well-established C18 columns needs to be conducted carefully. A positive affirmation about the SiGO-C18ec columns is that it is possible to confirm that it can be applied as a stationary phase and replace C18 columns occasionally. As a negative statement about the SiGO-C18ec columns, it is possible to affirm that it still demands several developments to be at a comparable stage with C18.

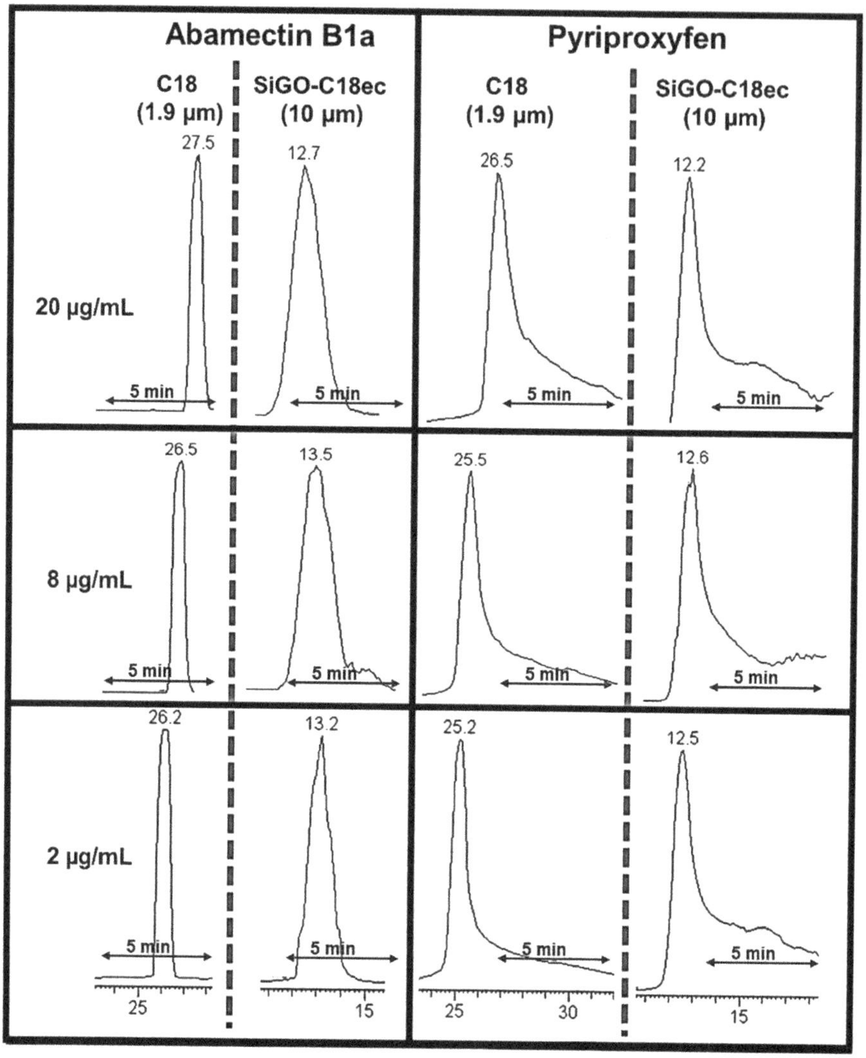

Figure 6. Comparisons of peak shape obtained for abamectin B1a and pyriproxyfen using C18 and SiGO-C18ec column as an analytical column in the column-switching system. Separation conditions: loading time of 3 min at 50 µL/min, linear gradient from 50% to 100% B in 15 min followed by an additional 5 min of 100% of B; 100 µL of analytes were injected.

The conclusions drawn via visual observation of peak shape were corroborated by analyzing the peak capacity performance parameter. In the proposed chromatographic method, the SiGO-C18ec column exhibited a peak capacity of six peaks in 15 min. In comparison, the C18 column demonstrated a peak capacity of 12 peaks under similar conditions, both at a concentration of 8 µg/mL. The peak capacity was determined at 13.4% of the peak height (equivalent to 4σ). However, the peak capacity observed in this study was lower than that reported in the literature for columns containing the same phase (SiGO-C18ec particles of 10 µm diameter). Notably, for hormone separation, a peak capacity of around 27 peaks in 15 min was reported [41], while for the separation of multiclass analytes, the reported peak capacity was 14 peaks in 15 min [42]. A comparison with other graphene-based stationary phases revealed similar outcomes. For instance, a peak capacity of approximately 19 peaks in 8 min was obtained to separate aromatic compounds using a stationary phase composed of octadecylsilane-functionalized graphene oxide coated over silica particles [37]. In another study, separations of anilines utilizing graphene oxide coated over silica particles displayed an estimated peak capacity of 15 peaks in 12 min [36]. Additionally, a significantly higher result, estimating a peak capacity of 57 peaks in 18 min, was reported for separating alkylbenzenes using a stationary phase composed of octadecyl amine and serine-derived carbon dot-modified silica gel [48]. The estimated peak capacity reported in the above three instances was evaluated based on chromatograms available in the references.

3. Material and Methods

3.1. Reagents and Standards

LC-grade acetone, acetonitrile, and MS-grade high-purity (>98%) formic acid were purchased from VWR International (Fontenay-sous-Bois, France). Deionized ultrapure water (H_2O) was produced in the lab by a Milli-Q® Millipore system (Burlington, NJ, USA). The SiGO and SiGO-C18ec fully porous particles (FPPs) were produced as described by Maciel et al. [41]. The C18 column packed with 1.9 µm superficially porous particles (SPPs) was a Hypersil GOLD 50 × 2.1 mm (Thermo Electronic Corporation, San Jose, CA, USA). Pyriproxyfen and abamectin B1a were acquired from Cymit Chimica (Barcelona, Spain). The orange-flavored carbonated soft drinks comprising 8% orange juice (Kas Naranja, PepsiCo, Spain) and 1.7% (Schweppes Naranja, Schweppes, S.A. Spain) were both acquired in a local market.

3.2. Sample Preparation

The analytes' standards were solubilized to a 1000 µg/mL solution in pure acetone and then diluted to 100 µg/mL in 90% acetonitrile and 10% acetone solution. The 100 µg/mL pesticide solution produced the 0.5–25 µg/mL samples by spiking the analytes in the matrix. The samples comprised 25% organic fraction and 75% orange-flavored carbonated soft drinks. The soft drink with 8% orange juice in the composition spiked with the analytes was used for the calibration curves, accuracy and precision, and comparison with C18 column experiments. Solutions of 2, 8, and 20 µg/mL of each orange-flavored carbonated soft drink and water (25% organic and 75% water) were used to evaluate the matrix effect. The orange-flavored carbonated soft drinks were selected as the matrix because, although it is a complex matrix, it can be directly injected into the system without pretreatment.

3.3. Analytical Instrumentation

A liquid chromatography Accela LC system composed of an auto-sampler, a column temperature controller, and a quaternary pump (Thermo Electronic Corporation, San Jose, CA, USA) coupled with a TSQ Quantum Access triple-quadrupole mass spectrometer (MS) fitted with an electrospray ion (ESI) source (Thermo Electronic Corporation, San Jose, CA, USA) was used in this work. A 100 µL loop was placed in the injection valve. Analyses were carried out using lab-made SiGO and SiGO-C18ec columns prepared as described by Borsatto et al. [42], using 100 × 0.3 mm tubing and 5 µm SiGO FPPs and 10 µm SiGO-C18ec FPPs. The column-switching system was assembled as shown in Figure 7. The SiGO column (extraction column) was connected between the injection valve and the MS valve, and the SiGO-C18ec column was connected between the MS valve and the ESI inlet.

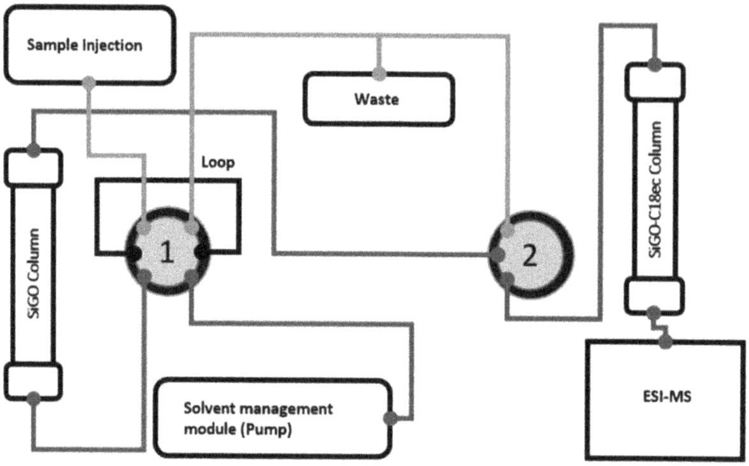

Figure 7. Scheme illustrating the column-switching system evaluated.

3.4. *Analytical Methods*

The mobile phase consisted of H_2O (A) and acetonitrile (B), both containing 0.1% of formic acid as an additive to enhance the positive electrospray (ESI) ionization process. In addition, a gradient method was used, as described in Table 2.

Table 2. HPLC column-switching events.

Event	Time	% B	Injection Valve Position	MS Valve Position
Pre-run	0 min	0%	Loading	Waste
Loading	0 to 3 min	0%	Inject	Waste
Gradient	3 to 18 min	50% to 100%	Inject	ESI-MS
Isocratic	18 to 23 min	100%	Inject	ESI-MS

An equilibration gradient drove the mobile phase composition back to 100% H_2O in 5 min, followed by an additional 10 min of constant 100% H_2O. The flow rate was kept constant at 50 µL/min, except for the loading flow rate experiment, in which 25 and 100 µL/min flows were also explored. The MS parameters were spray voltage of 4000 V, N_2 desolvation gas flow of 20 L/h, and capillary temperature of 400 °C. A multiple reaction monitoring (MRM) mode was used to observe the analytes. The quantifying transitions were 322 > 185 (collision energy of 20 V) for pyriproxyfen and 895 > 449 (collision energy of 40 V) for abamectin B1a. The qualifying transitions were 322 > 96 (collision energy of 40 V) for pyriproxyfen and 895 > 305 (collision energy of 70 V) for abamectin B1a.

4. Conclusions

This study investigated the applicability and limitations of a column-switching system composed exclusively of columns packed with graphene-based stationary phases. The samples of orange-flavored soft drinks with spiked analytes (25% of acetonitrile added to the samples) were directly injected without pretreatment or sample preparation. It was observed that the proposed column-switching system presented good linearity in the calibration curve and reasonable accuracy and precision (% RSD) for pyriproxyfen analysis, but not for abamectin B1a analysis. Furthermore, it demonstrated that this system could be applied to reliable quantitative analysis, but it is not multipurpose and presents limitations. The main restrictive factor was the limited capability of the SiGO extraction column to retain some analytes well. It was also observed that the proportion of orange juice in the matrix composition affected the results.

Additionally, it was noted that the life use of the SiGO extraction column was about 120 injections in the explored conditions, blockages and carry-over effects being observed. In the last stage of this study, the analytical SiGO-C18ec column was compared with a commercial C18 column. It was noted that although the C18 column presented better performance with thinner peaks and less tail, owing to the much smaller particle size of the stationary phase, the SiGO-C18ec columns can be applied for the same use with no practical restriction. In short, graphene-based particles are promising stationary phases for liquid chromatography and can be applied in both dimensions of a column-switching system. Although columns packed with these materials have been finding current applications in LC, improvements are still needed to increase the range of applications and life use. If these phases keep being developed and optimized, soon it might become an additional solution for separation science applications. Optimizations in the injection process and the detection method could improve the sensibility and applicability of this system proposed and might be an interesting topic to be investigated in the future. Furthermore, ramifications of this work to investigate different analytes in the various matrix could lead to a better understanding of these phases and help to draw the boundary conditions for these systems.

Author Contributions: Conceptualization, J.V.B.B.; methodology, J.V.B.B.; investigation, J.V.B.B. and E.V.S.M.; resources, F.M.L. and A.C.; writing—original draft preparation, J.V.B.B.; writing—review and editing, J.V.B.B., E.V.S.M., F.M.L. and A.C.; supervision, F.M.L. and A.C.; project administration, F.M.L. and A.C.; funding acquisition, F.M.L. and A.C. All authors have read and agreed to the published version of the manuscript.

Funding: The authors are grateful for the financial support from (i) the Coordination for the Improvement of Higher Education Personnel (CAPES—Brazil, Finance Code 001 and Proc. 88887.695324/2022-00); (ii) the São Paulo Research Foundation (FAPESP—Brazil, Grants 2019/22724-7, 2017/02147-0); (iii) the National Council for Scientific and Technological Development (CNPq—Brazil, 307293/2014-9; 308843/2019-3; Proc. 142513/2019-9); and (iv) the Spanish Ministry of Science and Innovation (PID2020-113050RB-I00 project).

Institutional Review Board Statement: Not applicable.

Informed Consent Statement: Not applicable.

Data Availability Statement: Data will be made available on request.

Acknowledgments: The Technical University of Darmstadt—Germany, and the Alexander von Humboldt Foundation—Germany contributed to support this work.

Conflicts of Interest: The authors declare that they have no known competing financial interests or personal relationships that could have appeared to influence the work reported in this paper.

Sample Availability: Not applicable.

References

1. Joshi, D.J.; Koduru, J.R.; Malek, N.I.; Hussain, C.M.; Kailasa, S.K. Surface Modifications and Analytical Applications of Graphene Oxide: A Review. *TrAC Trends Anal. Chem.* **2021**, *144*, 116448. [CrossRef]
2. Justino, C.I.L.; Gomes, A.R.; Freitas, A.C.; Duarte, A.C.; Rocha-Santos, T.A.P. Graphene Based Sensors and Biosensors. *TrAC Trends Anal. Chem.* **2017**, *91*, 53–66. [CrossRef]
3. Huang, T.; Tang, X.; Luo, K.; Wu, Y.; Hou, X.; Tang, S. An Overview of Graphene-Based Nanoadsorbent Materials for Environmental Contaminants Detection. *TrAC Trends Anal. Chem.* **2021**, *139*, 116255. [CrossRef]
4. Li, X.; Li, S.; Bai, J.; Peng, Y.; Ning, B.; Shi, H.; Kang, W.; Zhou, H.; Gao, Z. Determination of Bisphenol A by High-Performance Liquid Chromatography Based on Graphene Magnetic Dispersion Solid Phase Extraction. *J. Chromatogr. Sci.* **2020**, *58*, 280–286. [CrossRef]
5. Zhang, L.; Zhang, Y.; Tang, Y.; Li, X.; Zhang, X.; Li, C.; Xu, S. Magnetic Solid-Phase Extraction Based on Fe_3O_4/Graphene Oxide Nanoparticles for the Determination of Malachite Green and Crystal Violet in Environmental Water Samples by HPLC. *Int. J. Env. Anal. Chem.* **2018**, *98*, 215–228. [CrossRef]

6. Hou, X.; Ding, R.; Yan, S.; Zhao, H.; Yang, Q.; Wu, W. ZrO_2 Nanoparticles and Poly(Diallyldimethylammonium Chloride)-Doped Graphene Oxide Aerogel-Coated Stainless-Steel Mesh for the Effective Adsorption of Organophosphorus Pesticides. *Foods* **2021**, *10*, 1616. [CrossRef]
7. Sun, H.; Feng, J.; Feng, J.; Sun, M.; Feng, Y.; Sun, M. Carbon Aerogels Derived from Waste Paper for Pipette-Tip Solid-Phase Extraction of Triazole Fungicides in Tomato, Apple and Pear. *Food Chem.* **2022**, *395*, 133633. [CrossRef]
8. Gao, Y.; Gao, M.; Chen, G.; Tian, M.; Zhai, R.; Huang, X.; Xu, X.; Liu, G.; Xu, D. Facile Synthesis of Covalent Organic Frameworks Functionalized with Graphene Hydrogel for Effectively Extracting Organophosphorus Pesticides from Vegetables. *Food Chem.* **2021**, *352*, 129187. [CrossRef]
9. Ligorio, C.; Vijayaraghavan, A.; Hoyland, J.A.; Saiani, A. Acidic and Basic Self-Assembling Peptide and Peptide-Graphene Oxide Hydrogels: Characterisation and Effect on Encapsulated Nucleus Pulposus Cells. *Acta Biomater.* **2022**, *143*, 145–158. [CrossRef]
10. Wang, N.; Lu, Y.; Cui, B. Preparation and Application of β-Cyclodextrin Functionalised Graphene Oxide-Grafted Silica Sorbents for Solid-Phase Extraction (SPE) of Polycyclic Aromatic Hydrocarbons from Fried Food Using a Box-Behnken Design. *Food Anal. Methods* **2021**, *14*, 1577–1589. [CrossRef]
11. Song, X.-L.; Lv, H.; Wang, D.-D.; Liao, K.-C.; Wu, Y.-Y.; Li, G.-M.; Chen, Y. Graphene Oxide Composite Microspheres as a Novel Dispersive Solid-Phase Extraction Adsorbent of Bisphenols Prior to Their Quantitation by HPLC–Mass Spectrometry. *Microchem. J.* **2022**, *172*, 106920. [CrossRef]
12. Sun, M.; Han, S.; Maloko Loussala, H.; Feng, J.; Li, C.; Ji, X.; Feng, J.; Sun, H. Graphene Oxide-Functionalized Mesoporous Silica for Online in-Tube Solid-Phase Microextraction of Polycyclic Aromatic Hydrocarbons from Honey and Detection by High Performance Liquid Chromatography-Diode Array Detector. *Microchem. J.* **2021**, *166*, 106263. [CrossRef]
13. Pei, J.; Ren, T.; Huang, Y.; Chen, R.; Jin, W.; Shang, S.; Wang, J.; Liu, Z.; Liang, Y.; Abd El-Aty, A.M. Application of Graphene and Its Derivatives in Detecting Hazardous Substances in Food: A Comprehensive Review. *Front. Chem.* **2022**, *10*, 894759. [CrossRef]
14. Hu, H.; Wen, W.; Ou, J. Z Construction of Adsorbents with Graphene and Its Derivatives for Wastewater Treatment: A Review. *Environ. Sci. Nano* **2022**, *9*, 3226–3276. [CrossRef]
15. Li, Y.; Lan, S.; Zhu, T. Recent Advances of Graphene-Based Sorptive Materials in Extraction: A Review. *TrAC Trends Anal. Chem.* **2021**, *142*, 116319. [CrossRef]
16. Manousi, N.; Kabir, A.; Zachariadis, G.A. Recent Advances in the Extraction of Triazine Herbicides from Water Samples. *J. Sep. Sci.* **2022**, *45*, 113–133. [CrossRef]
17. Musarurwa, H.; Tawanda Tavengwa, N. Recent Progress in the Application of PH-Responsive Polymers in Separation Science. *Microchem. J.* **2022**, *179*, 107503. [CrossRef]
18. Jiang, X.; Ruan, G.; Huang, Y.; Chen, Z.; Yuan, H.; Du, F. Assembly and Application Advancement of Organic-functionalized Graphene-based Materials: A Review. *J. Sep. Sci.* **2020**, *43*, 1544–1557. [CrossRef] [PubMed]
19. Zhou, J.; Wang, R.; Chen, Z. Stir Bar Sorptive Extraction with a Graphene Oxide Framework-Functionalized Stainless-Steel Wire for the Determination of Sudan Dyes in Water Samples. *Anal. Methods* **2019**, *11*, 2050–2056. [CrossRef]
20. Wang, X.; Han, Y.; Cao, J.; Yan, H. Headspace Solid-Phase-Microextraction Using a Graphene Aerogel for Gas Chromatography–Tandem Mass Spectrometry Quantification of Polychlorinated Naphthalenes in Shrimp. *J. Chromatogr. A* **2022**, *1672*, 463012. [CrossRef] [PubMed]
21. Pei, M.; Shi, X.; Wu, J.; Huang, X. Graphene Reinforced Multiple Monolithic Fiber Solid-Phase Microextraction of Phenoxyacetic Acid Herbicides in Complex Samples. *Talanta* **2019**, *191*, 257–264. [CrossRef]
22. de Toffoli, A.L.; Fumes, B.H.; Lanças, F.M. Packed In-Tube Solid Phase Microextraction with Graphene Oxide Supported on Aminopropyl Silica: Determination of Target Triazines in Water Samples. *J. Env. Sci. Health B* **2018**, *53*, 434–440. [CrossRef]
23. Sun, R.; Lu, F.; Yu, C.; Yang, Y.; Qiao, L.; Liu, A. Peanut Shells-Derived Biochars as Adsorbents for the Pipette-Tip Solid-Phase Extraction of Endocrine-Disrupting Phenols in Water, Milk and Beverage. *J. Chromatogr. A* **2022**, *1673*, 463101. [CrossRef] [PubMed]
24. Grajek, H.; Jonik, J.; Witkiewicz, Z.; Wawer, T.; Purchała, M. Applications of Graphene and Its Derivatives in Chemical Analysis. *Crit. Rev. Anal. Chem.* **2020**, *50*, 445–471. [CrossRef]
25. Costa Queiroz, M.E.; Donizeti de Souza, I.; Marchioni, C. Current Advances and Applications of In-Tube Solid-Phase Microextraction. *TrAC Trends Anal. Chem.* **2019**, *111*, 261–278. [CrossRef]
26. Ponce-Rodríguez, H.D.; Verdú-Andrés, J.; Herráez-Hernández, R.; Campíns-Falcó, P. Innovations in Extractive Phases for In-Tube Solid-Phase Microextraction Coupled to Miniaturized Liquid Chromatography: A Critical Review. *Molecules* **2020**, *25*, 2460. [CrossRef]
27. Technical Analytical Methods Committee Briefs What Causes Most Errors in Chemical Analysis? *Anal. Methods* **2013**, *5*, 2914–2915. [CrossRef] [PubMed]
28. Liang, X.; Hou, X.; Chan, J.H.M.; Guo, Y.; Hilder, E.F. The Application of Graphene-Based Materials as Chromatographic Stationary Phases. *TrAC Trends Anal. Chem.* **2018**, *98*, 149–160. [CrossRef]
29. Lanças, F.M.; Medina, D.A.V.; Pereira Dos Santos, N.G.; Sinisterra, M.J. Graphene-Based Sorbents for Modern Magnetic Solid-Phase Extraction Techniques. In *Analytical Applications of Functionalized Magnetic Nanoparticles*; The Royal Society of Chemistry: London, UK, 2021; pp. 174–199.

30. dos Santos, N.G.P.; Maciel, E.V.S.; Mejía-Carmona, K.; Lanças, F.M. Multidimensional Capillary Liquid Chromatography-Tandem Mass Spectrometry for the Determination of Multiclass Pesticides in "Sugarcane Spirits" (Cachaças). *Anal. Bioanal. Chem.* **2020**, *412*, 7789–7797. [CrossRef]
31. Maclel, E.V.S.; Mejiá-Carmona, K.; Lancas, F.M. Evaluation of Two Fully Automated Setups for Mycotoxin Analysis Based on Online Extraction-Liquid Chromatography-TandemMass Spectrometry. *Molecules* **2020**, *25*, 2756. [CrossRef]
32. Mejía-Carmona, K.; Lanças, F.M. Modified Graphene-Silica as a Sorbent for in-Tube Solid-Phase Microextraction Coupled to Liquid Chromatography-Tandem Mass Spectrometry. Determination of Xanthines in Coffee Beverages. *J. Chromatogr. A* **2020**, *1621*, 461089. [CrossRef]
33. Maciel, E.V.S.; Vargas-Medina, D.A.; Lancas, F.M. Analyzes of β-Lactam Antibiotics by Direct Injection of Environmental Water Samples into a Functionalized Graphene Oxide-Silica Packed Capillary Extraction Column Online Coupled to Liquid Chromatography Tandem Mass Spectrometry. *Talanta Open* **2023**, *7*, 100185. [CrossRef]
34. Maciel, E.V.S.; de Toffoli, A.L.; da Silva Alves, J.; Lanças, F.M. Multidimensional Liquid Chromatography Employing a Graphene Oxide Capillary Column as the First Dimension: Determination of Antidepressant and Antiepileptic Drugs in Urine. *Molecules* **2020**, *25*, 1092. [CrossRef] [PubMed]
35. Liang, X.; Liu, S.; Song, X.; Zhu, Y.; Jiang, S. Layer-by-Layer Self-Assembled Graphene Oxide/Silica Microsphere Composites as Stationary Phase for High Performance Liquid Chromatography. *Analyst* **2012**, *137*, 5237. [CrossRef] [PubMed]
36. Zhang, X.; Chen, S.; Han, Q.; Ding, M. Preparation and Retention Mechanism Study of Graphene and Graphene Oxide Bonded Silica Microspheres as Stationary Phases for High Performance Liquid Chromatography. *J. Chromatogr. A* **2013**, *1307*, 135–143. [CrossRef] [PubMed]
37. Liang, X.; Wang, S.; Liu, S.; Liu, X.; Jiang, S. A Novel Octadecylsilane Functionalized Graphene Oxide/Silica Composite Stationary Phase for High Performance Liquid Chromatography. *J. Sep. Sci.* **2012**, *35*, 2003–2009. [CrossRef] [PubMed]
38. Borsatto, J.V.B.; Maciel, E.V.S.; Cifuentes, A.; Lanças, F.M. Online Extraction Followed by LC–MS/MS Analysis of Lipids in Natural Samples: A Proof-of-Concept Profiling Lecithin in Seeds. *Foods* **2023**, *12*, 281. [CrossRef]
39. Liang, X.; Wang, X.; Ren, H.; Jiang, S.; Wang, L.; Liu, S. Gold Nanoparticle Decorated Graphene Oxide/Silica Composite Stationary Phase for High-Performance Liquid Chromatography. *J. Sep. Sci.* **2014**, *37*, 1371–1379. [CrossRef]
40. Li, Y.; Li, Q.; Zhu, N.; Gao, Z.; Ma, Y. Cellulose Type Chiral Stationary Phase Based on Reduced Graphene Oxide@silica Gel for the Enantiomer Separation of Chiral Compounds. *Chirality* **2018**, *30*, 996–1004. [CrossRef]
41. Maciel, E.V.S.; Borsatto, J.V.B.; Mejia-Carmona, K.; Lanças, F.M. Application of an In-House Packed Octadecylsilica-Functionalized Graphene Oxide Column for Capillary Liquid Chromatography Analysis of Hormones in Urine Samples. *Anal. Chim. Acta* **2023**, *1239*, 340718. [CrossRef]
42. Borsatto, J.V.B.; Maciel, E.V.S.; Lanças, F.M. Investigation of the Applicability of Silica-Graphene Hybrid Materials as Stationary Phases for Capillary Liquid Chromatography. *J. Chromatogr. A* **2022**, *1685*, 463618. [CrossRef] [PubMed]
43. Ferrer, C.; Martínez-Bueno, M.J.; Lozano, A.; Fernández-Alba, A.R. Pesticide Residue Analysis of Fruit Juices by LC–MS/MS Direct Injection. One Year Pilot Survey. *Talanta* **2011**, *83*, 1552–1561. [CrossRef] [PubMed]
44. by Liquid Chromatography–Electrospray Tandem Mass Spectrometry. *J. Chromatogr. A* **2003**, *992*, 133–140. [CrossRef]
45. Lee, T.D. Introduction to Modern Liquid Chromatography, Third Edition. *J. Am. Soc. Mass. Spectrom.* **2011**, *22*, 196. [CrossRef]
46. González-Ruiz, V.; Olives, A.I.; Martín, M.A. Core-Shell Particles Lead the Way to Renewing High-Performance Liquid Chromatography. *TrAC Trends Anal. Chem.* **2015**, *64*, 17–28. [CrossRef]
47. Blue, L.E.; Jorgenson, J.W. 1.1μm Superficially Porous Particles for Liquid Chromatography. Part II: Column Packing and Chromatographic Performance. *J. Chromatogr. A* **2015**, *1380*, 71–80. [CrossRef]
48. Zhao, C.; Yan, S.; Liu, J.; Xiong, Z.; Zhao, L. Octadecylamine and Serine-Derived Carbon Dots-Modified Silica Gel for Reversed Phase/Hydrophilic Interaction Liquid Chromatography. *Microchem. J.* **2022**, *183*, 107987. [CrossRef]

molecules

MDPI

Article

Water Dynamics in Starch Based Confectionery Products including Different Types of Sugar

Esmanur İlhan [1], Pelin Poçan [2], Danuta Kruk [3,*], Miłosz Wojciechowski [4], Maciej Osuch [3], Roksana Markiewicz [5], Stefan Jurga [5] and Mecit Halil Oztop [1]

1 Department of Food Engineering, Middle East Technical University, Ankara 06800, Turkey; esmanur@metu.edu.tr (E.İ.); mecit@metu.edu.tr (M.H.O.)
2 Department of Food Engineering, Faculty of Engineering and Architecture, Konya Food and Agriculture University, Konya 42080, Turkey; pelin.pocan@gidatarim.edu.tr
3 Department of Physics & Biophysics, Faculty of Food Sciences, University of Warmia and Mazury in Olsztyn, Michala Oczapowskiego 4, 10-719 Olsztyn, Poland; maciej.osuch@uwm.edu.pl
4 Faculty of Mathematics and Computer Science, University of Warmia and Mazury in Olsztyn, Sloneczna 54, 10-710 Olsztyn, Poland; wojciechowski@matman.uwm.edu.pl
5 NanoBioMedical Centre, Adam Mickiewicz University, Wszechnicy Piastowskiej 3, 61-614 Poznan, Poland; roksana.markiewicz@amu.edu.pl (R.M.); stjurga@amu.edu.pl (S.J.)
* Correspondence: danuta.kruk@uwm.edu.pl

Abstract: Starch-based confectionery products were prepared using different types of sugar. In addition to using different sugar, starch was replaced with soy protein isolate (SPI) in some of the products. ^{1}H NMR spin-lattice relaxation experiments were performed for the collection of products in a broad frequency range from 4 KHz to 30 MHz to get insight into the influence of different sugar types and SPI on the dynamics of water in composite gel systems. The relaxation data have been decomposed into relaxation contributions associated with two different pools of water molecules characterized by different mobility. The translation dynamics of water molecules has been quantitatively described in terms of a dedicated relaxation model. The influence of the sample composition (the type of sugar and/or the presence of SPI) on the water mobility was thoroughly discussed. The results indicate that the addition of soy protein does not affect water dynamics for samples including sucrose. In addition, as the complementary measurements, physical properties of the products, such as the moisture content, water activity and texture, were investigated in terms of X-ray diffraction and thermogravimetric analysis.

Keywords: fast field cycling (FFC) NMR relaxometry; confectionery; starch; soy protein isolate; D-allulose; relaxation; dynamics

Citation: İlhan, E.; Poçan, P.; Kruk, D.; Wojciechowski, M.; Osuch, M.; Markiewicz, R.; Jurga, S.; Oztop, M.H. Water Dynamics in Starch Based Confectionery Products including Different Types of Sugar. *Molecules* **2022**, *27*, 2216. https://doi.org/10.3390/molecules27072216

Academic Editor: Gavino Sanna

Received: 7 March 2022
Accepted: 25 March 2022
Published: 29 March 2022

1. Introduction

Confectionery gels are composed of high amounts of sugar components such as sucrose and glucose syrup, gelling agents such as starch, gelatin, or pectin along with food flavorings and colorings. In non-confectionery gel systems, gelation involves the dissolution of biopolymers in an aqueous environment and subsequent gelation via a crosslinking agent [1,2]. However, the addition of sugar greatly influences the standard gelation mechanism due to the low mobility of water and high solid fraction [3].

The high sucrose and glucose syrup content of confectionary gels is an increasing concern due to high sugar intake. Therefore, reducing sugar consumption leads one to use low and non-calorie sweeteners as a substitute for sucrose. Recently, it has been shown that the use of rare sugar, D-Allulose (a type of monosaccharide found in nature in small amounts), has yielded desirable characteristics in terms of processing and rheological properties of the confectionary products [4–6] with positive health effects [7–10]. D-allulose (C-3 epimer of fructose with a ketone group) has 70% of the sweetness of sucrose with a lower caloric value of 0.39 kcal/g due to poor digestion [11,12]. It has been shown that D-allulose exhibits different water-binding properties compared to other monosaccharides [13], and physical properties of the confectionery products can be affected by the water binding ability of sugars used in formulations. D-allulose has a wide array of biomedical applications, such as improving insulin resistance, its anti-obesity effects, its anti-inflammatory nature, and in regulating glucolipid metabolism [14,15]. It is also worth mentioning that D-allulose intake improves cholesterol metabolism, leading thus to a reduced risk of atherosclerotic plaque formation, which is considered a major cause of ischemic heart disease [16].

Proteins and polysaccharides are widely used as components of gel matrices. Depending on relative ratios of these components and the concentration of the exceeding polymer, one can observe segregation or complexation effects [17–19]. Therefore, a proper combination and concentration of polymers is essential for the formation of gel matrices and their stability. This subject becomes even more complex for confectionery gel systems containing several gelling agents, including proteins and polysaccharides [5,20–25].

In this work, we focused on the influence of D-allulose and soy protein isolate on the dynamic properties of the starch-based composite gel matrices and the relationship between the dynamical (molecular) and macroscopic features was exploited by nuclear magnetic resonance (NMR) relaxometry.

There is a growing interest in food science in applying NMR relaxometry as a tool enabling the linking of macroscopic properties of food products with dynamics on the molecular level. NMR relaxometry has been applied to enquire into the dynamical properties of several kinds of food, including eggs [26], where differences in water dynamics in different kinds of eggs and the influence of storage on dynamical properties of water were investigated. Whey-protein-based composite hydrogels [27], with different formulations, were examined to understand the dynamics of water molecules enclosed in the systems. One should also mention the works devoted to the aging of banana and spoilage of milk [28], the determination of the dynamics of virgin rape oil molecules [29], the characterization of balsamic vinegars of different aging processes [30], and the establishment of a relationship between translational diffusion coefficients and the viscosity of different kinds of oil for controlling the authenticity of oil products [31]. Furthermore, as far as sugar-containing products are considered, NMR relaxometry has recently been applied to investigate molecular properties of gelatin-based soft candies [32] in which the macroscopic properties of gelatin-based confections were correlated with water performance and provided methodological guidance on how to use the FFC NMR relaxometry to obtain a quantitative characterization of these products.

In the present work, the influence of D-allulose (replacing sucrose) and soy protein isolate on water mobility in starch-based confectionery products has been investigated by means of FFC NMR relaxometry. In addition, as a complementary method to FFC NMR relaxometry, thermal gravimetric analysis experiments (TGA), water activity, moisture content, and X-ray diffraction (XRD) analysis were also conducted, and the results were compared accordingly.

2. Results

2.1. NMR Relaxometry

The ^{1}H spin-lattice relaxation data obtained for the set of samples listed in Table 1 are presented in Figure 1. Two observations can be made at this stage—the first one is that the data for the two samples of each kind ((1) and (2)) are in a good agreement. Moreover, the other replicates were also consistent with each other. The second observation is that the relaxation rates for 9S30 and 11S30 are very close, while the relaxation rates for 9_R30 and 11_R30 differ from them and are considerably different between each other.

Table 1. Parameters obtained from the analysis of the relaxation data in terms of Equation (2).

Sample	C^{s}_{DD} [Hz^2]	τ^{s}_{c} [s]	C^{f}_{DD} [Hz^2]	τ^{f}_{c} [s]	A [s^{-1}]
9_R30 (1)	$1.06 \times 10^{9} \pm 1.02 \times 10^{7}$	$1.61 \times 10^{7} \pm 3.67 \times 10^{-8}$	$1.67 \times 10^{9} \pm 4.55 \times 10^{7}$	$1.97 \times 10^{-8} \pm 2.90 \times 10^{-9}$	13.1
9_R30 (2)	$1.10 \times 10^{9} \pm 1.43 \times 10^{7}$	$1.69 \times 10^{-7} \pm 5.18 \times 10^{-8}$	$1.67 \times 10^{9} \pm 6.23 \times 10^{7}$	$2.06 \times 10^{-8} \pm 4.23 \times 10^{-9}$	13.1
9_S30 (1)	$1.14 \times 10^{9} \pm 1.25 \times 10^{7}$	$1.43 \times 10^{-7} \pm 4.02 \times 10^{-8}$	$1.29 \times 10^{9} \pm 1.32 \times 10^{8}$	$1.32 \times 10^{-8} \pm 4.70 \times 10^{-9}$	10.0
9_S30 (2)	$1.12 \times 10^{9} \pm 1.01 \times 10^{7}$	$1.45 \times 10^{-7} \pm 3.33 \times 10^{-8}$	$1.30 \times 10^{9} \pm 9.72 \times 10^{7}$	$1.37 \times 10^{-8} \pm 3.81 \times 10^{-9}$	10.0
11_R30 (1)	$4.71 \times 10^{8} \pm 6.84 \times 10^{6}$	$1.32 \times 10^{-7} \pm 4.52 \times 10^{-8}$	$1.42 \times 10^{9} \pm 1.84 \times 10^{7}$	$1.88 \times 10^{-8} \pm 1.86 \times 10^{-9}$	16.5
11_R30 (2)	$4.84 \times 10^{8} \pm 6.91 \times 10^{6}$	$1.31 \times 10^{-7} \pm 4.45 \times 10^{-8}$	$1.46 \times 10^{9} \pm 1.79 \times 10^{7}$	$1.86 \times 10^{-8} \pm 1.84 \times 10^{-9}$	15.5
11_S30 (1)	$1.12 \times 10^{9} \pm 1.25 \times 10^{7}$	$1.50 \times 10^{-7} \pm 4.35 \times 10^{-8}$	$1.36 \times 10^{9} \pm 1.38 \times 10^{8}$	$1.33 \times 10^{-8} \pm 4.81 \times 10^{-9}$	9.0
11_S30 (2)	$1.11 \times 10^{9} \pm 1.13 \times 10^{7}$	$1.50 \times 10^{-7} \pm 3.89 \times 10^{-8}$	$1.31 \times 10^{9} \pm 1.10 \times 10^{8}$	$1.40 \times 10^{-8} \pm 4.36 \times 10^{-9}$	10.1

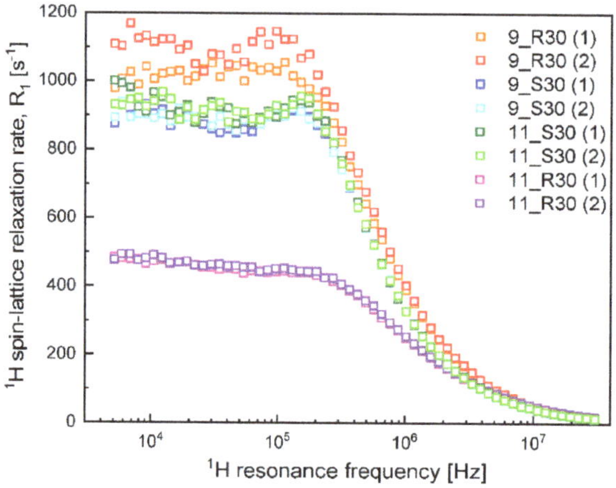

Figure 1. ^{1}H spin-lattice relaxation data for the prepared samples.

^{1}H relaxation processes are caused by magnetic dipole-dipole interactions. The interactions fluctuate in time as a result of molecular motion. As the relaxation experiments have been carried out in the frequency range encompassing five decades, the relaxation rates are associated with dynamical processes occurring on considerably different time scales—at low frequencies, one observes slow dynamics, while with increasing frequency a progressively faster dynamics is probed. In the simplest case of a single dynamical process contributing to the relaxation, the relaxation rate, $R_1(\omega)$ (ω) denotes the resonance frequency in angular frequency units), can be expressed as [27];

$$R_1(\omega)=C_{DD}\left[\frac{\tau_c}{1+(\omega\tau_c)^2}+\frac{4\tau_c}{1+(2\omega\tau_c)^2}\right] \tag{1}$$

where τ_c denotes a characteristic time constant of the dynamical process, referred to as a correlation time, while C_{DD} is the corresponding dipolar relaxation constant. Anticipating the results, it has turned out that the relaxation data of Figure 1 can be reproduced in terms of two dynamical processes and a frequency independent relaxation contribution:

$$R_1(\omega)=C^{s}_{DD}\left[\frac{\tau^{s}_{c}}{1+(\omega\tau^{s}_{c})^2}+\frac{4\tau^{s}_{c}}{1+(2\omega\tau^{s}_{c})^2}\right]+C^{f}_{DD}\left[\frac{\tau^{f}_{c}}{1+(\omega\tau^{f}_{c})^2}+\frac{4\tau^{f}_{c}}{1+(2\omega\tau^{f}_{c})^2}\right]+A \tag{2}$$

The pairs of parameters τ^{s}_{c}, C^{s}_{DD} and τ^{f}_{c}, C^{f}_{DD} denote the correlation time and the corresponding dipolar relaxation constant for the slower and the faster dynamical processes, respectively. The frequency independent term, A, represents a relaxation contribution associated, in fact, with a motion that is so fast that its correlation time, τ_c, fulfills the condition: $\omega\tau_c \ll 1$ and, consequently, the corresponding expression does not depend on frequency (in the frequency range exploited in this work).

The results of the analysis of the data in terms of Equation (2) are shown in Figure 2; the theoretical curves have been decomposed into the individual relaxation contributions.

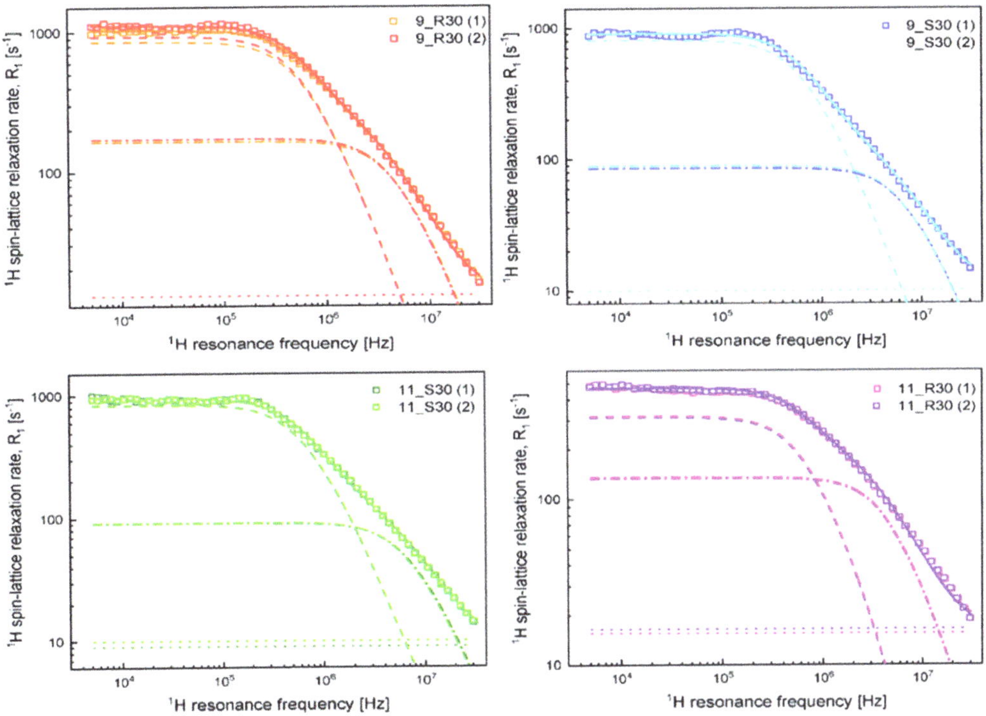

Figure 2. ^{1}H spin-lattice relaxation data for samples listed in Table 3. Solid lines—fits in terms of Equation (2) decomposed into the relaxation contributions associated with the slow dynamics (dashed lines), fast dynamics (dashed-dotted lines), and the frequency independent term (dotted lines).

The obtained parameters are collected in Table 1.

2.2. *Water Activity, Moisture Content, Hardness, Thermogravimetric Analysis, and X-ray Diffraction*

In addition to the NMR relaxation studies, water activity, moisture content and hardness for the different gel formulations have been measured. The results are collected in Table 2. The moisture content is described by the percentage of the ratio of the weight of water to the total weight of the material. The hardness values (N) represent peak forces during compression of the samples, while the temperatures at which the derivative off the mass loss show peaks (minima) are referred to as peak temperatures (°C).

Table 2. Water activity (a_w), moisture content (MC %), hardness (N) and peak temperatures (°C) for different formulations.

Sample Name	aw	MC %	Hardness (N)	Peak Temperature (°C)
9_R30	0.54 ± 0.00^{d}	12.19 ± 0.00^{c}	06.07 ± 0.57^{b}	133.68 ± 0.71^{ab}
9_S30	0.66 ± 0.00^{b}	14.09 ± 0.01^{b}	23.92 ± 0.81^{a}	123.15 ± 3.08^{bc}
11_R30	0.61 ± 0.01^{c}	13.85 ± 0.07^{b}	02.58 ± 0.24^{c}	141.07 ± 0.49^{a}
11_S30	0.69 ± 0.00^{a}	14.94 ± 0.01^{a}	06.08 ± 0.11^{b}	119.75 ± 5.72^{c}

Means within the same column, followed by the different small letters are significantly different for each sample ($p < 0.05$).

The measurements have been complemented by thermogravimetric analysis (TGA). The results are shown in Figure 3a,b, while the peak temperatures are included in Table 2.

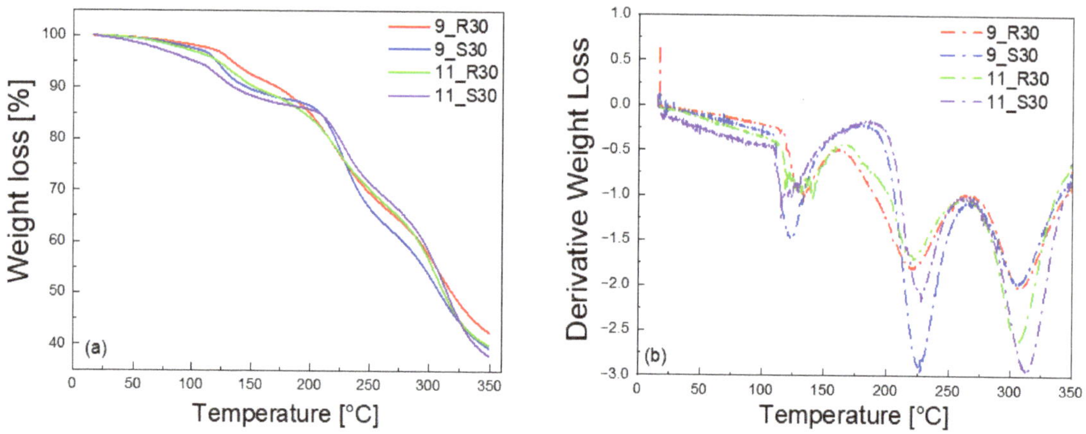

Figure 3. TGA results for 9_R30, 9_S30, 11_R30, and 11_S30 (**a**); derivatives of the mass loss versus temperature (**b**).

To complete the characterization of the samples, X-ray diffraction (XRD) experiments have been performed. The XRD patterns are shown in Figure 4.

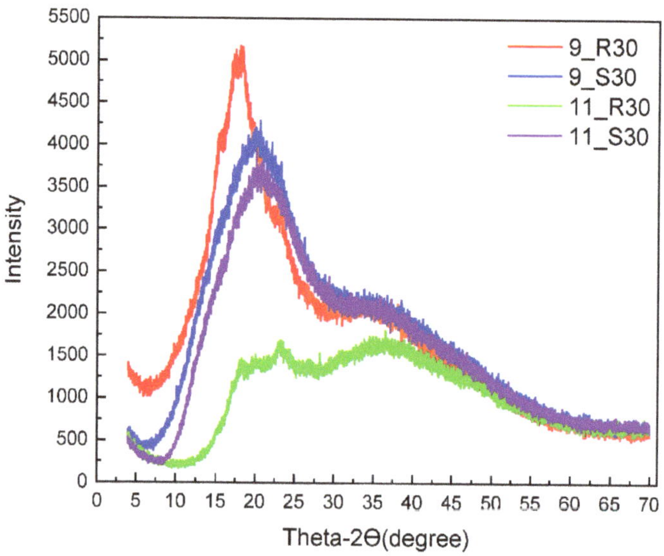

Figure 4. XRD patterns for 9_R30, 9_S30, 11_R30 and 11_S30.

3. Discussion

One can clearly see from Figure 1 that at low frequencies, the ^{1}H spin-lattice relaxation rates for 9_R30 are higher than for 9_S30—then with increasing frequency, the relaxation rates tend to coincide (in the high frequency range). The relaxation data have been interpreted in terms of a model assuming the presence of two (at least) pools of water molecules characterized by different mobility. The dynamics of both fractions is considerably slowed down by the confinement (interactions with the sugar fraction), nevertheless, to distinguish between them, we use the terminology "slow" and "fast". The characteristic correlation time τ_c^s for the "slow" fraction of water molecules yields = 1.65×10^{-7} s for 9_R30, while the corresponding dipolar relaxation constant C_{DD}^s yields 1.08×10^9 Hz^2. Analogous values for 9_S30 are given as $\tau_c^s = 1.44 \times 10^{-7}$ s and $C_{DD}^s = 1.13 \times 10^9$ Hz^2. This comparison shows that the parameters characterizing the "slow" fraction of water in 9_R30 and 9_S30 are similar. The correlation times for the "fast" water fraction is equal to $\tau_c^f = 2.03 \times 10^{-7}$ s for 9_R30 and $\tau_c^f = 1.35 \times 10^{-7}$ s for 9_S30. This implies that dynamics of the "fast" water fraction for 9_R30 is slower than for 9_S30. It is worth noting at this stage that the dynamics of the "slow" fraction is in a good approximation slower by an order of magnitude compared to the dynamics of the "fast" water fraction. The dipolar relaxation constant C_{DD}^f for 9_R30 is 1.67×10^9 Hz^2, while for 9_S30 it yields 1.30×10^9 Hz^2. Dipolar relaxation constants include (are proportional to) the factor Pq, where P is the mole faction of water in the bound position, while q denotes the coordination number [26]. This implies that the product Pq for the "fast" water fraction is larger for 9_R30 than for 9_S30.

Following this line, the relaxation data (and, consequently, the parameters) for 11_S30 are very similar to those for 9S_30: for the "slow" water fraction in 11_S30, it has been obtained: $\tau_c^s = 1.50 \times 10^{-7}$ s, $C_{DD}^s = 1.12 \times 10^9$ Hz^2, while for the "fast" water fraction, the parameters yield: $\tau_c^f = 1.36 \times 10^{-8}$ s, $C_{DD}^f = 1.34 \times 10^9$ Hz^2. The results show that the addition of soy protein (2%) does not affect the water dynamics as long as the samples include sucrose. When the sucrose is replaced by D-allulose in the soy protein containing samples, the relaxation rates become much smaller (Figure 1), and the corresponding parameters show significant differences. Comparing the parameters for 11_S30 with those for 11_R30, one sees that while the correlation time $\tau_c^s = 1.32 \times 10^{-7}$ s for 11_R30 for the "slow" water fraction is close to the corresponding value for 11_S30, the dipolar relaxation constant for 11_R30, $C_{DD}^s = 4.78 \times 10^8$ Hz^2, is lower by more than la factor of two compared to the corresponding value for 11_S30. As far as the "fast" water fraction is concerned, the dipolar relaxation constant for 11_R30, $C_{DD}^f = 1.44 \times 10^9$ Hz^2 is close to the value for 11S-30, while the correlation time $\tau_c^f = 1.87 \times 10^{-8}$ s for 11R_30 is somewhat longer than for 11S_30. As the dipolar relaxation constant for the "slow" water fraction in 11_R30 is also by more than a factor of two lower than the corresponding value for 9R_30, one can conclude that the mole fraction of bound water molecules for 11_R30 is by more than a factor of two lower than for 9_R30 (both samples include D-allulose, so the coordination number, q, is the same for both cases) and the difference is solely caused by the presence of soy protein in the small amount of 2%. One should also comment about the frequency independent term, C. This relaxation contribution represents a dynamical process that occurs on a fast time scale—of the order of a couple of ns or faster. Consequently, as the condition: $\omega\tau_c \ll 1$ (where τ_c denotes the corresponding correlation time) is fulfilled, in that case contribution becomes frequency independent. This relaxation term likely originates from several processes: dynamics of a "free" water fraction (a fraction of water molecules, the dynamics of which is not so significantly affected by interactions with the sugar and protein molecules) and/or dynamics of functional groups of sugar molecules. We would prefer not to speculate on this subject, but it is worthy of note that the A term is higher in the presence of D-allulose.

The D-allulose containing samples (9_R30 and 11_R30) are characterized by lower water activity and moisture content compared to their sucrose containing counterparts. In [5], it has been stated that the water-binding ability of D-allulose is lower than that of sucrose. Consequently, it has been argued that during the preparation of the sugar syrup mixture, D-allulose-containing formulations might have lost more water due to evaporation, which could have resulted in lower "free" water and moisture content in the final product [5]. Independently of these considerations, the water activity and moisture content vary in rather narrow ranges that clearly indicates that these quantities are not affected by the population of the "slow" water fraction.

One should point out the relationship between the relaxation properties and the XRD patterns for the four formulations. XRD experiments were performed to understand the crystallinity of the confectionary formulations. The highest and the most narrow XRD peak has been observed for 9_R30—this corresponds to the fastest relaxation process (fast relaxation is characteristic of materials of high crystallinity). Then the XRD patterns for 9_S30 and 11_S30 almost overlap— and so do the relaxation data. Eventually the broadest XRD peak is observed for 11_R30 (indicating a distribution of crystals, i.e., polydispersity of the sample), corresponding to the slowest relaxation reflected by the relatively small population of the "slow" water fraction.

The presence of soy proteins increases the hardness independently of the sugar (D-allulose or sucrose)—this is clearly seen from Table 3 (6.07 for 9_R30 versus 2.58 for 11_R30 and 23.92 for 9_S30 versus 6.08 for 11_S30). At the same time, one observes that hardness decreases upon replacing sucrose by D-allulose (23.92 for 9_S30 versus 6.07 for 9_R30 and 6.08 for 11_S30 versus 2.58 for 11_R30) independently of the presence of soy proteins. Amine groups present in protein chains react with glucose and D-Allulose upon heat treatment through Maillard reaction. The addition of soy proteins to the formulation means introducing amine groups that react with the monosaccharides. The reaction results in the formation of covalent cross-links within the protein network [33] that might be the reason for the increase of hardness in the presence of soy proteins.

Table 3. Composition of the formulations.

Name	Starch (%)	D-Allulose (%)	Sucrose (%)	Corn Syrup (%)	Soy Protein (%)
9_R30	9	30	0	30	2
9_S30	9	0	30	30	2
11_R30	11	30	0	30	0
11_S30	11	0	30	30	0

A closer inspection of the TGA curves and their derivatives indicates a three-step mass loss for all samples in the temperature range from 25 °C to 350 °C. The mass loss in the first (low temperature) stage can be attributed to losing water [34], while the high temperature ones can be associated with the decomposition of polymers and organic compounds in the formulations. The temperature position of the first minimum in the derivative curves e (starting from the low temperature) reveals that for sucrose-containing samples (11_S30 and 9_S30), water escapes from the system at 119 °C and 123 °C, respectively. On the other hand, for the D-allulose containing samples (11_R30 and 9_R30), water escapes from the system at 141 °C and 133 °C, which are slightly higher compared to the sucrose containing samples.

4. Materials and Methods

4.1. Sample Preparation

Four kinds of starch-based gels were prepared, and are shown in Table 3. The first pair (9_R30 and 9_S30) includes 9% starch and 2% soy protein—the difference is in replacing the sucrose (present in 9_S30) by D-allulose (9_R30). The second pair (11_R30 and 11_S30) includes 11% starch (no soy protein) and again glucose (11_S30) and D-allulose (11_R30). For each case, two samples were prepared. In all cases, the remaining contribution (29%) is water.

The gels were prepared according to the method described in [5]. The amount of water used before cooking was 29% of the total mass weight. The water amount was divided into two to gelatinize the starch and to prepare the sugar solutions. In the first part, starch was mixed with water in the proportion of 1:2 (starch: water) and gelatinized in an oil bath at 140 °C for 5 min. The sugar syrup—powder sugar mixture was mixed in a glass beaker with the remaining water and boiled up to 115 °C. After that, the gelatinized starch was mixed with the syrup mixture at 115 °C. The soy protein isolate was added at this stage (for the formulations including soy protein isolate) and homogenized at 10,000 rpm for 1 min (WiseTisHG-15D homogenizer, Wertheim, Germany). The cooking continued in an oil bath at 140 °C until the mixture attained 75°Brix value. oBrix values of the formulations were measured by a hand refractometer (Hanna, HI96801, Smithfield, RI, USA). The hot mixture was then poured into starch molds with dimensions of 2.5 × 2.5 × 2 cm and kept at 38 °C for 36 h.

4.2. Experimental Methods

The ^{1}H spin-lattice relaxation experiments have been carried out in the frequency range from 5 kHz to 30 MHz using a STELAR (Mede, Italy) Spinmaster relaxometer. The experiments have been performed at 25 °C with a temperature accuracy of 1 °C. For each resonance frequency, 16 points have been collected for the magnetization curve (^{1}H magnetization versus time). The relaxation process has turned out to be single-exponential; the corresponding magnetization curves are presented in Supplementary Material. Consequently, the 1H spin-lattice relaxation rates have been obtained from single-exponential fits of the magnetization versus time curves for each resonance frequency.

Thermogravimetric analysis (TGA) was performed with a Perkin Elmer Pyris1 (Perkin Elmer, MA, USA) in the temperature range from 25 °C to 350 °C with a heating rate of 5 °C/min, under nitrogen.

X-ray diffraction experiments were performed with a Rigaku Ultima-IV X-Ray Diffractometer (Japan). The data were collected by the method of [35] in the range of 4–70 °C.

Hardness of the starch-based gels was measured by using a Texture Analyzer (Lloyd Instruments, TA Plus, Hants, UK). A 35-mm cylinder shape probe of a diameter of 1 cm and load cell of 50 N was attached to the instrument. The samples were compressed twice with 100 mm/min pretest speed. For the data analysis, NEXIGEN texture analysis software was used.

Moisture content of the samples was measured at 70 °C for 4 h in a vacuum oven (DAIHAN, Wonju, Korea). Weight loss of the samples was recorded, and the moisture content of each sample was calculated on that basis.

An Aqualab 4 TE (METER Group, Pullman, WA, USA) was used to measure the water activities of the samples. The experiments were conducted at 25 °C.

5. Conclusions

^{1}H spin-lattice relaxation data of starch-based, soy protein containing composite gel systems were collected in a wide frequency range (4 kHz to 30 MHz) and quantitatively analyzed. Two water fractions referred as the slow-water fraction and the fast-water fraction have been identified by finding the characteristic correlation time and the corresponding dipolar relaxation constant. The results show that the addition of soy protein (2%) does not affect the water dynamics as long as the samples include sucrose. When the sucrose is replaced by D-allulose in the soy protein containing samples, the relaxation rates become much smaller, and the corresponding parameters show significant differences. As the dipolar relaxation constant for the "slow" water fraction in D-allulose containing samples, the mole fraction of bound water molecules is lower by more than a factor of two for those where only starch is used as a gelling agent.

In this study, the unique potential of NMR relaxometry was exploited to get insight into water dynamics in the gel network. It should be pointed out that NMR relaxometry is a very useful method to investigate the properties of composite gels that are very common in food systems.

Supplementary Materials: The following supporting information can be downloaded at: https://www.mdpi.com/article/10.3390/molecules27072216/s1, Figures S1–S8: Normalized magnetization curves.

Author Contributions: Conceptualization, E.İ. and M.H.O.; Formal analysis, D.K. and M.O.; Funding acquisition, D.K., S.J. and M.H.O.; Investigation, E.İ. and M.H.O.; Methodology, E.İ., P.P., D.K. and M.W.; Project administration, D.K., S.J. and M.H.O.; Resources, D.K. and M.H.O.; Software, D.K. and M.O.; Supervision, D.K. and M.H.O.; Validation, E.İ., D.K. and M.O.; Visualization, E.İ., D.K. and M.O.; Writing—original draft, E.İ., P.P., D.K. and M.H.O.; Writing—review & editing, E.İ., P.P., D.K.,

M.W., M.O., R.M., S.J. and M.H.O. All authors have read and agreed to the published version of the manuscript.

Funding: Starch candy formulation and physical measurements sections has been funded by The Scientific and Technological Research Council of Turkey with the Grant # 116O759 under COST Program (EURELAX). From the Polish Side the work has been supported by National Science Centre, Poland, grant number: 2015/19/B/NZ9/03348. R.M. and S.J. would like to acknowledge the National Centre for Research and Development, Poland, grant number DWM/WPC2/285/2020.

Institutional Review Board Statement: Not applicable.

Informed Consent Statement: Not applicable.

Data Availability Statement: Not applicable.

Acknowledgments: Authors would like to acknowledge EURELAX COST Action (CA 15209).

Conflicts of Interest: The authors declare that they have no conflict of interest.

Sample Availability: Samples of the compounds are available from the authors.

References

1. Burey, P.; Bhandari, B.R.; Rutgers, R.P.G.; Halley, P.J.; Torley, P.J. Confectionery Gels: A Review on Formulation, Rheological and Structural Aspects. *Int. J. Food Prop.* **2009**, *12*, 176–210. [CrossRef]
2. Tolstoguzov, V. Phase behaviour of macromolecular components in biological and food systems. *Food/Nahrung* **2000**, *44*, 299–308. [CrossRef]
3. Kasapis, S.; Al-Marhoobi, I.M.; Deszczynski, M.; Mitchell, J.R.; Abeysekera, R. Gelatin vs Polysaccharide in Mixture with Sugar. *Biomacromolecules* **2003**, *4*, 1142–1149. [CrossRef] [PubMed]
4. Poçan, P.; Ilhan, E.; Oztop, M.H. Effect of D-psicose substitution on gelatin based soft candies: A TD-NMR study. *Magn. Reson. Chem.* **2019**, *57*, 661–673. [CrossRef] [PubMed]
5. Ilhan, E.; Poçan, P.; Ogawa, M.; Oztop, M.H. Role of 'D-allulose' in a starch based composite gel matrix. *Carbohydr. Polym.* **2020**, *228*, 115373. [CrossRef] [PubMed]
6. O'Charoen, S.; Hayakawa, S.; Matsumoto, Y.; Ogawa, M. Effect of D-Psicose Used as Sucrose Replacer on the Characteristics of Meringue. *J. Food Sci.* **2014**, *79*, E2463–E2469. [CrossRef] [PubMed]
7. Zeng, Y.; Zhang, X.; Guan, Y.; Sun, Y. Characteristics and Antioxidant Activity of Maillard Reaction Products from Psicose-Lysine and Fructose-Lysine Model Systems. *J. Food Sci.* **2011**, *76*, C398–C403. [CrossRef]
8. Sun, Y.; Hayakawa, S.; Ogawa, M.; Izumori, K. Antioxidant properties of custard pudding dessert containing rare hexose, d-psicose. *Food Control* **2007**, *18*, 220–227. [CrossRef]
9. Chung, M.-Y.; Oh, D.-K.; Lee, K.W. Hypoglycemic Health Benefits of D-Psicose. *J. Agric. Food Chem.* **2012**, *60*, 863–869. [CrossRef] [PubMed]
10. Matsuo, T.; Izumori, K. Effects of Dietary D-Psicose on Diurnal Variation in Plasma Glucose and Insulin Concentrations of Rats. *Biosci. Biotechnol. Biochem.* **2006**, *70*, 2081–2085. [CrossRef] [PubMed]
11. Granström, T.B.; Takata, G.; Tokuda, M.; Izumori, K. Izumoring: A novel and complete strategy for bioproduction of rare sugars. *J. Biosci. Bioeng.* **2004**, *97*, 89–94. [CrossRef]
12. Mu, W.; Zhang, W.; Feng, Y.; Jiang, B.; Zhou, L. Recent advances on applications and biotechnological production of D-psicose. *Appl. Microbiol. Biotechnol.* **2012**, *94*, 1461–1467. [CrossRef] [PubMed]
13. Ikeda, S.; Gohtani, S.; Fukada, K.; Amo, Y. Dielectric Relaxation and Water Activity in Aqueous Solution of D-Psicose. *Jpn. J. Food Eng.* **2011**, *12*, 67–74. [CrossRef]
14. Xia, Y.; Cheng, Q.; Mu, W.; Hu, X.; Sun, Z.; Qiu, Y.; Liu, X.; Wang, Z. Research Advances of D-allulose: An Overview of Physiological Functions, Enzymatic Biotransformation Technologies, and Production Processes. *Foods* **2021**, *10*, 2186. [CrossRef] [PubMed]
15. Kanasaki, A.; Jiang, Z.; Mizokami, T.; Shirouchi, B.; Iida, T.; Nagata, Y.; Sato, M. Dietary d-allulose alters cholesterol metabolism in Golden Syrian hamsters partly by reducing serum PCSK9 levels. *J. Funct. Foods* **2019**, *60*, 103429. [CrossRef]
16. Marzilli, M.; Merz, C.N.B.; Boden, W.E.; Bonow, R.O.; Capozza, P.G.; Chilian, W.M.; DeMaria, A.N.; Guarini, G.; Huqi, A.; Morrone, D.; et al. Obstructive Coronary Atherosclerosis and Ischemic Heart Disease: An Elusive Link! *J. Am. Coll. Cardiol.* **2012**, *60*, 951–956. [CrossRef]
17. de Kruif, C.; Tuinier, R. Polysaccharide protein interactions. *Food Hydrocoll.* **2001**, *15*, 555–563. [CrossRef]
18. Grinberg, V.; Tolstoguzov, V. Thermodynamic incompatibility of proteins and polysaccharides in solutions. *Food Hydrocoll.* **1997**, *11*, 145–158. [CrossRef]
19. Tolstoguzov, V.B. Protein-polysaccharide interactions. *Food Proteins Appl.* **2017**, *5*, 171–198. [CrossRef]
20. Ates, E.G.; Ozvural, E.B.; Oztop, M.H. Understanding the role of D-Allulose and soy protein addition in pectin gels. *J. Appl. Polym. Sci.* **2021**, *138*, 49885. [CrossRef]

21. Gu, J.; Ahn-Jarvis, J.H.; Vodovotz, Y. Development and Characterization of Different Black Raspberry Confection Matrices Designed for Delivery of Phytochemicals. *J. Food Sci.* **2015**, *80*, E610–E618. [CrossRef] [PubMed]
22. Marfil, P.H.M.U.; Anhê, A.C.B.M.; Telis, V.R.N.U. Texture and Microstructure of Gelatin/Corn Starch-Based Gummy Confections. *Food Biophys.* **2012**, *7*, 236–243. [CrossRef]
23. Ong, M.H.; Whitehouse, A.; Abeysekera, R.; Al-Ruqaie, I.M.; Kasapis, S. Glass transition-related or crystalline forms in the structural properties of gelatin/oxidised starch/glucose syrup mixtures. *Food Hydrocoll.* **1998**, *12*, 273–281. [CrossRef]
24. Sessler, T.; Weiss, J.; Vodovotz, Y. Influence of pH and soy protein isolate addition on the physicochemical properties of functional grape pectin confections. *Food Hydrocoll.* **2013**, *32*, 294–302. [CrossRef]
25. Siegwein, A.M.; Vodovotz, Y.; Fisher, E.L. Concentration of Soy Protein Isolate Affects Starch-Based Confections' Texture, Sensory, and Storage Properties. *J. Food Sci.* **2011**, *76*, E422–E428. [CrossRef]
26. Kruk, D.; Florek-Wojciechowska, M.; Oztop, M.; Ilhan, E.; Wieczorek, Z. Water dynamics in eggs by means of Nuclear Magnetic Resonance relaxometry. *J. Magn. Reson.* **2021**, *327*, 106976. [CrossRef]
27. Ozel, B.; Kruk, D.; Wojciechowski, M.; Osuch, M.; Oztop, M.H. Water Dynamics in Whey-Protein-Based Composite Hydrogels by Means of NMR Relaxometry. *Int. J. Mol. Sci.* **2021**, *22*, 9672. [CrossRef] [PubMed]
28. Steele, R.M.; Korb, J.-P.; Ferrante, G.; Bubici, S. New applications and perspectives of fast field cycling NMR relaxometry. *Magn. Reson. Chem.* **2015**, *54*, 502–509. [CrossRef] [PubMed]
29. Rachocki, A.; Tritt-Goc, J. Novel application of NMR relaxometry in studies of diffusion in virgin rape oil. *Food Chem.* **2014**, *152*, 94–99. [CrossRef]
30. Baroni, S.; Consonni, R.; Ferrante, G.; Aime, S. Relaxometric Studies for Food Characterization: The Case of Balsamic and Traditional Balsamic Vinegars. *J. Agric. Food Chem.* **2009**, *57*, 3028–3032. [CrossRef] [PubMed]
31. Kruk, D.; Masiewicz, E.; Budny, J.; Stankiewicz, A.; Lotarska, S.; Oztop, M.; Wieczorek, Z. Diffusion in oils versus their viscosity —Insight from Nuclear Magnetic Resonance relaxometry. *J. Food Eng.* **2022**, *317*, 110848. [CrossRef]
32. Pocan, P.; Ilhan, E.; Florek–Wojciechowska, M.; Masiewicz, E.; Kruk, D.; Oztop, M.H. Exploring the water mobility in gelatin based soft candies by means of Fast Field Cycling (FFC) Nuclear Magnetic Resonance relaxometry. *J. Food Eng.* **2021**, *294*, 110422. [CrossRef]
33. Sun, Y.; Hayakawa, S.; Izumori, K. Modification of Ovalbumin with a Rare Ketohexose through the Maillard Reaction: Effect on Protein Structure and Gel Properties. *J. Agric. Food Chem.* **2004**, *52*, 1293–1299. [CrossRef] [PubMed]
34. Fisher, E.L.; Ahn-Jarvis, J.; Gu, J.; Weghorst, C.M.; Vodovotz, Y. Assessment of physicochemical properties, dissolution kinetics and storage stability of a novel strawberry confection designed for delivery of chemopreventive agents. *Food Struct.* **2014**, *1*, 171–181. [CrossRef]
35. Suput, D.Z.; Lazic, V.L.; Popovic, S.Z.; Hromis, N.M. Edible films and coatings: Sources, properties and application. *Food Feed Res.* **2015**, *42*, 11–22. [CrossRef]

molecules

MDPI

Article

Development of a Gas-Tight Syringe Headspace GC-FID Method for the Detection of Ethanol, and a Description of the Legal and Practical Framework for Its Analysis, in Samples of English and Welsh Motorists' Blood and Urine

Luke Taylor †, Lili Saskőy †, Tara Brodie, Vytautas Remeškevičius, Hannah Jayne Moir, James Barker, John Fletcher, Baljit Kaur Thatti, Gavin Trotter and Brian Rooney *

Faculty of Science, Engineering and Computing, School of Life Sciences, Pharmacy and Chemistry, Kingston University, Kingston upon Thames, London KT1 2EE, UK; l.taylor@kingston.ac.uk (L.T.); k1700497@kingston.ac.uk (L.S.); t.brodie@kingston.ac.uk (T.B.); k1501503@kingston.ac.uk (V.R.); h.moir@kingston.ac.uk (H.J.M.); j.barker@kingston.ac.uk (J.B.); j.fletcher@kingston.ac.uk (J.F.); b.thatti@kingston.ac.uk (B.K.T.); g.trotter@kingston.ac.uk (G.T.)

* Correspondence: b.rooney@kingston.ac.uk

† These authors contributed equally to this work.

Citation: Taylor, L.; Saskőy, L.; Brodie, T.; Remeškevičius, V.; Moir, H.J.; Barker, J.; Fletcher, J.; Kaur Thatti, B.; Trotter, G.; Rooney, B. Development of a Gas-Tight Syringe Headspace GC-FID Method for the Detection of Ethanol, and a Description of the Legal and Practical Framework for Its Analysis, in Samples of English and Welsh Motorists' Blood and Urine. *Molecules* **2022**, *27*, 4771. https://doi.org/10.3390/molecules27154771

Academic Editor: Gavino Sanna

Received: 17 June 2022
Accepted: 17 July 2022
Published: 26 July 2022

Abstract: Ethanol is the most commonly used recreational drug worldwide. This study describes the development and validation of a headspace gas chromatography flame ionisation detection (HS-GC-FID) method using dual columns and detectors for simultaneous separation and quantitation. The use of a dual-column, dual-detector HS-GC-FID to quantitate ethanol is a common analytical technique in forensic toxicology; however, most analytical systems utilise pressure-balance injection rather than a simplified gas-tight syringe, as per this technique. This study is the first to develop and validate a technique that meets the specifications of the United Kingdom's requirements for road traffic toxicology testing using a Shimadzu GC-2014 gas-tight syringe. The calibration ranged from 10 to 400 mg/100 mL, with a target minimum linearity of $r2 > 0.999$, using tertiary butanol as the internal standard marker. The method has an expanded uncertainty at 99.73% confidence of 3.64% at 80 mg/100 mL, which is the blood alcohol limit for drink driving in England and Wales. In addition, at 200 mg%—the limit at which a custodial sentence may be imposed on the defendant—the expanded uncertainty was 1.95%. For both the 80 mg% and 200 mg% concentrations, no bias was present in the analytical method. This method displays sufficient separation for other alcohols, such as methanol, isopropanol, acetaldehyde, and acetone. The validation of this technique complies with the recommended laboratory guidelines set out by United Kingdom and Ireland Association of Forensic Toxicologists (UKIAFT), the recently issued Laboratory 51 guidelines by the United Kingdom Accreditation Service (UKAS), and the criteria set out by the California Code of Regulations (CCR), 17 CCR § 1220.1.

Keywords: HS-GC-FID; ethanol; method validation; gas-tight syringe; method validation

1. Introduction

Ethanol is the most encountered recreational drug in forensic toxicology [1,2]. Due to its depressant effects on the central nervous system (CNS), ethanol causes significant impairment of driving performance [3]. Most countries have set alcohol concentration limits for drivers. In England and Wales, the drink driving limit in blood is set at 80 mg/100 mL (0.08%), with equivalent urine and breath limits of 107 mg/100 mL and 35 µg/100 mL, respectively. These limits are among the highest in the world. By contrast, Scotland and many European countries have set a 50 mg/100 mL ethanol limit in blood [4]. Northern Ireland has also proposed an amendment to the law to lower the prosecution limit to

50 mg/100 mL in blood, along with an additional limit of 20 mg/100 mL for newly qualified drivers and taxi drivers, which is yet to be accepted at the time of publication [5].

In England and Wales, motorists found above the prescribed limit are charged under Section 5 of the Road Traffic Act 1988 [6]. This includes driving, attempting to drive, or being in charge of a motor vehicle on a road or other public place after consuming so much alcohol that the proportion of it in the breath, blood, or urine exceeds the prescribed limit. While most drink driving cases are prosecuted on the basis of evidential breath readings, there is still a requirement for blood or urine analysis in cases in which a breath sample cannot be obtained [7]. To aid law enforcement, rapid, cost-effective, and accurate methods to quantify ethanol in blood and urine are needed. Headspace gas chromatography with flame ionisation detection (GC-FID) is the industry standard alcohol analysis technique, due to its effectiveness at detecting volatile compounds with minimal sample preparation [8,9].

Headspace gas chromatography operates on the principle of Henry's law—that the concentration ratio of a volatile substance in the gas phase of a closed vessel and in the liquid phase is fixed at a given pressure and temperature. There are a number of different techniques for transferring a sample from the headspace to the chromatography columns [10,11]. These include filled-loop, pressure-balance [12], and direct injection systems [13,14].

The use of headspace gas chromatography for the analysis of ethanol in bodily fluids is not new. Many such methods have been published in the past [15,16]. However, there are differences in approach and regulations between countries, as well as differences in the equipment used. The US state of California regulates road traffic alcohol analysis through its Code of Regulations (CCR), 17 CCR § 1220.1 [17], while in the UK there are no specific regulations relating to this analysis.

PerkinElmer's pressure-balance headspace sampling system is widely used in the UK for testing road traffic ethanol samples. We "inherited" a Shimadzu gas-tight syringe sampling system—a Shimadzu GC-2014 gas chromatograph equipped with an HTA 200 H headspace autosampler. The main interest of this research was to see how this sampling technique would perform in comparison to the pressure-balance system. Since we did not have access to the pressure-balance system, a direct comparison could not be made, so we assessed the gas-tight syringe system against the California codes and the de facto quality criteria used in England and Wales. The California codes were chosen arbitrarily, as California is a sizeable jurisdiction outside of the UK whose codes we had access to.

2. Results

2.1. Separation Conditions

The California codes, 1220.1 (a) (2), require that the method be capable of the analysis of ethanol with a specificity that is adequate and appropriate for traffic law enforcement. Figure 1 details the specificity of both columns, and demonstrates that the BAC plus 2 column showed complete separation of ethanol from any of the common interferences that might be encountered in road traffic cases. BAC plus 1 did not entirely separate ethanol from acetone; however, concentrations of acetone of 10 mg/100 mL and below did not interfere with ethanol resolution and/or quantification. It is our understanding that in California the results of one column are used for quantitation, with the results of the other column being used to confirm the identification of ethanol. In England and Wales, the results of both columns are used for quantitation. The precision of the four sub-results (duplicate analysis on dual columns) is checked. Typically, limits of either a %CV of 2.5, or an absolute limit of within +/−2 mg/100 mL (0.002%), are applied. In cases where this limit is not met, due to the presence of interference from acetone, the analysis is repeated, and quantitation is undertaken using the four sub-results from the BAC plus 2 column.

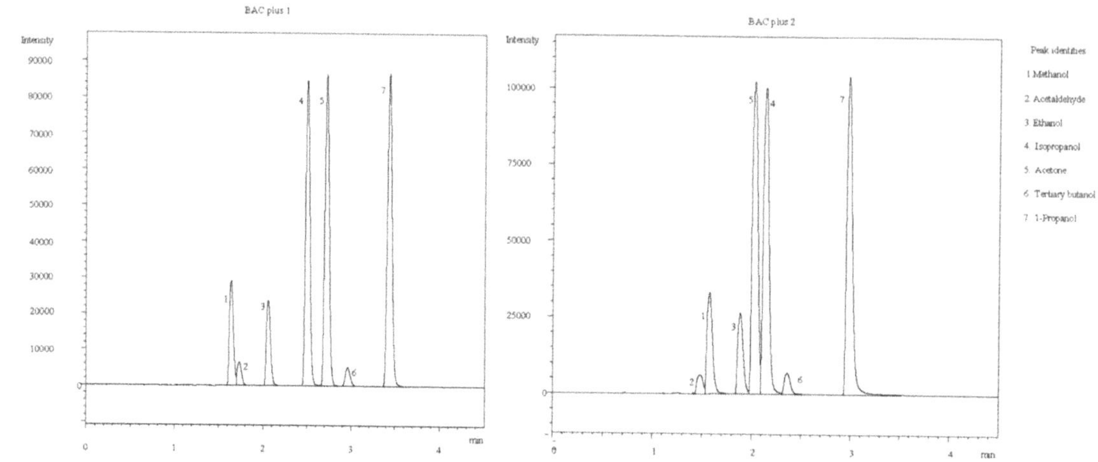

Figure 1. Separation of ethanol from commonly detected alcohols: Both columns demonstrated sufficient separation of ethanol from a range of alcohols, including methanol, isopropanol, tertiary butanol, and 1-propanol, in addition to ethanol's major metabolite acetaldehyde. Separation of ethanol and acetone was achieved on BAC 1, while BAC 2 demonstrated co-elution with acetone at concentrations greater than 10 mg/mL.

2.2. Linearity

Linearity was assessed using calibrators at 10, 20, 50, 100, 200, and 400 mg/100 mL (Table 1). In England and Wales, a minimum permitted R^2 value of 0.998 is typically applied in road traffic alcohol cases. During validation, 11 separate extractions were analysed; in all analyses the measured R^2 value in both the BAC 1 and 2 columns was greater than 0.999. In addition, the average residual sum of squares was found to be 0.00336 and 0.00339 for BAC 1 and BAC 2, respectively.

Table 1. Coefficients of determination for dual-column, dual-detector GC-FID analysis of ethanol: The R^2 values on both the BAC 1 (column A) and BAC 2 (column B) columns in all validation batches exceeded the minimum requirement of 0.998 for road traffic toxicology testing.

	Linearity	
Batch Number	**Linearity Column A**	**Linearity Column B**
1	0.99995	0.99995
2	0.99998	0.99998
3	0.99997	0.99995
4	0.99998	0.99995
5	0.99976	0.99974
6	0.99997	0.99998
7	0.99972	0.99975
8	0.99963	0.99983
9	0.99962	0.99997
10	0.99991	0.99993
11	0.99986	0.99986
Average	0.99985	0.99990
Standard deviation	1.42059×10^{-4}	9.09551×10^{-5}
CV%	0.01421	0.00910

2.3. Precision

The intra-run and inter-run precision was assessed using ANOVA. No significant differences were found for the quality controls at 80 and 200 mg/100 mL, so the intermediate precision was calculated as the standard deviation of the set of duplicate results from the 11 batches. Significant differences were found at the value of 20 mg/100 mL, so the intermediate precision was calculated as the square root of the sum of squares of the intra- and inter-run precision. The California codes stipulate a precision limit of +/−5% of the value for alcohol concentrations above 0.100% (100 mg/100 mL) and +/−0.005% (5 mg/100 mL) for alcohol concentrations lower than 0.100%. The precision of our method was found to be +/−0.0006% at 200 mg/dL, +/−0.0012% at 80 mg/100 mL, and +/−0.0031% at 20 mg/100 mL (Table 2).

Table 2. Summary of the precision and accuracy of the validation results from 11 batches in mg/100 mL. No significant differences were found at 80 and 200 mg/100 mL—the two limits critical in road traffic toxicology casework.

	Precision and Accuracy		
	20 (mg/100 mL)	**80 (mg/100 mL)**	**200 (mg/100 mL)**
Mean	20.889	79.520	198.836
Precision	0.612	1.125	3.104
Accuracy	0.892	0.480	1.164

2.4. Accuracy

The California codes stipulate an accuracy limit of +/−5% of the value for alcohol concentrations above 0.100% (100 mg/100 mL) and +/−0.005% (5 mg/100 mL) for alcohol concentrations lower than 0.100%. The accuracy of our method was found to be +/−0.00116% at 200 mg/100 mL, +/−0.00048% at 80 mg/100 mL, and +/−0.00089% of the value at 20 mg/100 mL (Table 2). The bias was evaluated using a *t*-test incorporating the uncertainty of the target—the certified reference material uncertainty—into the measurement uncertainty. The bias was found not to be significant for the quality control samples at 80 mg/100 mL and 200 mg/100 mL. There was significant bias for the control samples at 20 mg/100 mL. The method uncertainty was calculated by combining the method intermediate precision with the uncertainty of the bias measurement as the square root of the sum of the squares.

2.5. Uncertainty

The method has a standard uncertainty at 200 mg/100 mL of 0.65%, at 80 mg/100 mL of 1.2%, and at 20 mg/100 mL of 1.77%, giving expanded uncertainties of 1.95% (or 3.88 mg/100 mL), 3.64% (or 2.89 mg/100 mL), and 5.31% (or 1.11 mg/100 mL), respectively, at a 99.73% confidence. In England and Wales, 6%, or 6 mg/100 mL—whichever is greater—is subtracted from the measured value as an allowance for measurement uncertainty. The result is then reported as being "not less than" this subtracted figure.

Figure 2 shows the quality control charts of the three different QC levels from 10 batches run on different days in duplicate analysis, once the method validation had been completed. The lower and upper warning and control limits in this method were defined as +/−2 and +/−3 standard deviations of the mean of validation QC values. The California codes recommend the acceptable limits to be set as +/−10 mg/100 mL of the mean value determined from 20 replicate analyses.

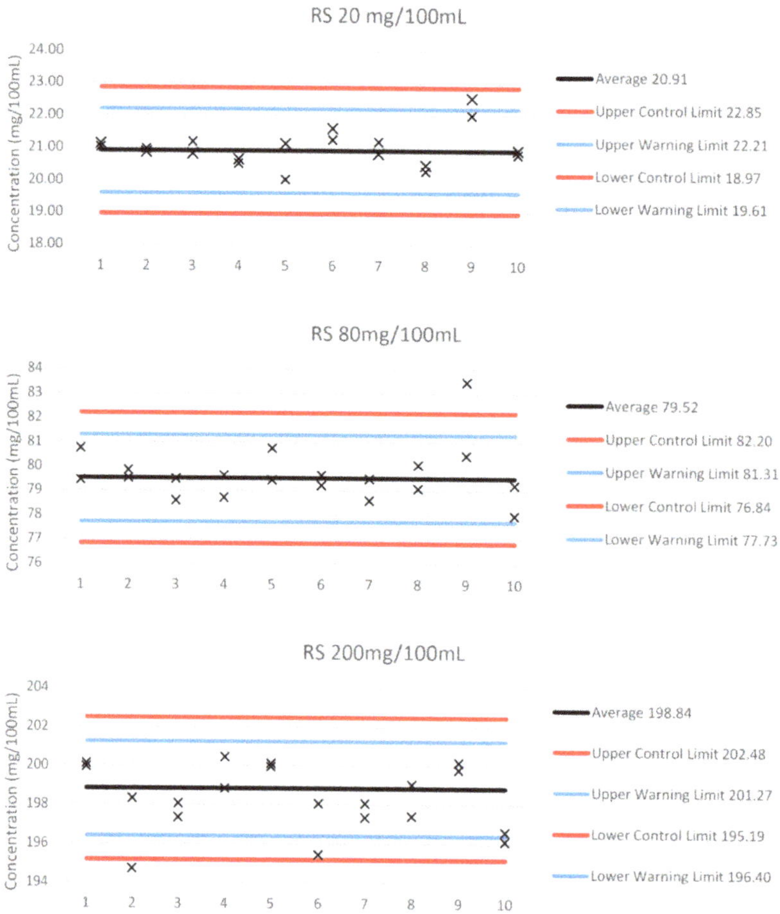

Figure 2. Determination of method uncertainty using the certified reference material: The method has a standard uncertainty at 200 mg/100 mL of 0.65%, at 80 mg/100 mL of 1.2%, and at 20 mg/100 mL of 1.77%, giving expanded uncertainties of 1.95% (or 3.88 mg/100 mL), 3.64% (or 2.89 mg/100 mL), and 5.31% (or 1.11 mg/100 mL), respectively, at a 99.73% confidence.

2.6. Limit of Detection and Quantification

Ethanol was qualitatively detected at concentrations of 5 mg/100 mL with a signal-to-noise ratio greater than 3:1—concentrations approximately half that of the limit of quantitation. Qualitative assessment of the 5 mg/100 mL concentrations from four separate extractions on separate days showed an average ethanol concentration of 6.1 mg/100 mL (detected below the lowest calibrator) and an average ethanol peak area of 2162. The limit of quantitation (LOQ) was set at the lowest calibrator sample—10 mg/100 mL. This methodology has been deployed in range of sample types, including research studies [18], proficiency test trials, and road traffic casework analysis. From the analysis of 30 blood samples analysed over a period of approximately 18 months, the range of blood alcohol concentrations was 6.55 to 230.97 mg/100 mL, the average standard deviation was 1.55, the average %CV value was 2.96, and the median concentration was 43.44 mg/100 mL. Representative chromatograms are detailed in Supplementary Figure S1.

3. Discussion

This study details a gas-tight syringe GC-FID method that is capable of quantifying ethanol in blood samples taken in road traffic cases. There is no specific regulation of road traffic alcohol analysis in England and Wales, save for the requirement for methods used by prosecution laboratories to be ISO-17,025-accredited. To varying degrees, laboratories use methods and quality acceptance criteria based on those of the now-closed Forensic Science Service. Our methodology meets the general laboratory criteria as described by the United Kingdom and Ireland Association of Forensic Toxicologists (UKIAFT) [19] and the California Code of Regulations for the quantification of ethanol in road traffic samples.

This analytical technique utilises two separate columns and detectors; the columns chosen were the Restek BAC-1 and BAC-2, and each analytical result is the mean of the four sub-results arising from duplicate analysis on dual columns. The spread of those sub-results was limited to 2.5% at and above 80 mg/100 mL, and 2.5 mg/100 mL below 80. The method uses sodium metabisulphite as an antioxidant, but does not use a salting-out agent. The volume of internal standard added to each 100 µL sample is 1 mL; this floods the sample and prevents salting-out that may occur at lower volumes.

To the best of our knowledge, this publication describes the first dual-column/detector gas-tight syringe HS-GC-FID method that can accurately quantitate ethanol concentrations in blood samples to a forensic standard. Other studies have utilised the GC-FID pressure-balance system [20–22]. Our results compare favourably with a 2020 study by Mihretu et al., who detailed a HS-GC-FID technique to detect BAC using a PerkinElmer Clarus 500 pressure-balance injection. This technique had a correlation coefficient of $r^2 = 0.993$, and a precision (repeatability) and intermediate precision of 27% and 11%, respectively. This technique also displayed a limit of detection of 9.9 mg/100 mL, and did not use dual-column analysis—a requirement for the forensic testing of road traffic alcohol samples [20].

A correlation coefficient (r^2) limit of 0.9980 was applied to the linear calibration curves. The limit of quantitation (LOQ)—the lower end of the calibration—was 10 mg/100 mL (0.01%). This method utilises three different concentrations of quality controls: 20 (0.020%), 80 (0.080%), and 200 mg/100 mL (0.200%). The 20 mg/100 mL concentration is used to assess the lower blood alcohol limit for toxicology cases involving aviation staff, and to ensure that quality is maintained below the legal limit should back-calculations (retrograde calculations) be required. The QC at 200 mg/100 mL is used to demonstrate the method's performance above the 80 mg/100 mL legal limit, as increased sanctions are applied to motorists convicted of driving with BACs significantly above that limit. At present, the magistrates' court sentencing guidelines [23] specify a fine and a minimum driving ban of 12 months for BACs between 81 and 137 mg/100 mL (0.081 to 0.137%) (the offence is exceeding the limit of 80). A fine and a minimum ban of 17 months are imposed at BACs between 138 and 206 mg/100 mL (0.138 to 0.206%), while a community order and a minimum ban of 23 months are imposed at BACs between 207 and 275 mg/100 mL. A custodial sentence of up to 12 weeks may be imposed in addition to a minimum 29-month ban for BACs between 276 and 345 mg/100 mL (0.276 and 0.345%) and above. The starting sanctions are increased for a range of conditions, including prior convictions and aggravating factors such as carrying passengers.

The QCs are assessed using QC charts with limits based on the method precision, rather than the blanket +/−10 mg/100 mL (0.010) specified in the California codes. Those limits are shown in Figure 2. The California codes require mean QC values to be determined as the mean of at least 20 replicate analyses, at a rate of no more than two analyses per day. In the UK, no such specification exists. We assessed the QCs over 11 batches, with samples run in duplicate, as this is the specification set for Road Traffic Act Section 5 A: drug-driving. The 200 mg/100 mL QC chart shows that the limits are too narrow. Even the sorts of batch numbers specified by the California codes, along with the requirements for Section 5 A, may be insufficient to properly assess the method's performance. The precision calculated during validation is smaller than it appears to be after validation. We intend to re-assess the QC charts, along with other data, once sufficient data have been collected.

Carryover and matrix interferences were assessed by running blank internal standard samples after the highest calibrator and between duplicate samples and quality controls in every batch. No carryover was observed under the conditions described above. This complies with the requirements for laboratory testing set out by UKIAFT and the California codes.

Results in England and Wales are reported as a subtracted "not less" than value: 6 mg/100 mL, or 6%—whichever is higher—is subtracted from the analytical result. An analytical result of 87.15 mg/100 mL, for example, would be reported as "the sample contains alcohol at a concentration of not less than 81 mg/100 mL". This subtraction is an allowance for analytical uncertainty. This common approach is used by all laboratories in the UK, and ensures consistency between laboratories. The value of 6 mg/100 mL or 6% is based on the precision of the FSS analytical method introduced in the early 1970s. It far exceeds the uncertainty of modern methods. In the present study, the combined expanded uncertainty (at 99.73%) of our method at 80 mg/100 mL was found to be 2.89 mg/100 mL. In 2010, Sir Peter North recommended the use of 3 mg/100 mL or 3%—whichever is higher—as the subtraction that should be made to allow for analytical uncertainty. However, without regulation, and with the 6 mg or 6% over 100 mg subtraction present in case law, there is no obvious way for this to be implemented.

4. Materials and Methods

4.1. Chemicals and Reagents

Ethanol standard solutions were purchased from Cerilliant at concentrations of 10, 20, 50, 100, 200, and 400 mg/100 mL (Austin, TX, USA). The certified reference material—aqueous ethanol—was purchased from LGC (Teddington, UK) at concentrations of 20, 80, and 200 mg/100 mL. Anhydrous tertiary butanol and sodium metabisulphite were purchased from Fisher Scientific (Red Rock, CO, USA). Propan-1-ol, propan-2-ol, methanol, acetaldehyde, and acetone were obtained from Fisher Scientific (Red Rock, CO, USA)—all analytical grade. Deionised water was generated on-site using a Sartorius Arium™ advance EDI Water System, ASTM Type 2. Defibrinated horse blood was obtained from TCS Biosciences LTD (Buckingham, UK). Human blood samples were collected on-site from volunteers. The sample collection involving human participants was in accordance with the institutional research committee and the 1964 Helsinki Declaration, or comparable ethical standards. Furthermore, informed consent was obtained from all individuals who donated their samples for testing for either research or casework testing. The research samples collected were approved by the Faculty Research Ethics Committee (FREC) of Kingston University London (ethics code: 1819063.1) Proficiency testing samples were sourced from LGC's Forensic Blood Toxicology QUARTZ scheme.

4.2. Instrumentation

The headspace was sampled using an HTA 200 H headspace autosampler fitted with a gas-tight syringe. The chromatographic analysis was carried out on a Shimadzu GC-2014 gas chromatography system equipped with two FIDs. RTx-BAC 1 (30 m, 0.32 mm ID, 1.8 um) and RTx-BAC 2 (30 m, 0.32 mm ID, 0.6 um) columns (Restek, Bellefonte, PA, USA) were connected to the inlet using a Universal Y tight-press connector. Helium was used as the carrier gas at a column flow rate of 2.78 mL/min. Hydrogen was used as the FID fuel source, compressed air was used as the FID flame oxygen supply, and nitrogen was used as the makeup gas. During method development, investigations into the optimal conditions for efficient and thorough separation of ethanol, tert-butanol, and acetaldehyde were prioritised. The method optimisation involved the modification of GC oven temperature, pressure, and sample conditioning time. Our investigations assessed method performance with a range of pressures from 150 to 65 kPa; in addition, we also researched the effects of modifications in oven temperature between 40 and 50 °C and sample incubation times of 5, 10, 15, and 20 min. Our results indicated that the optimal parameters selected for this method were 40 °C GC oven temperature, a 5-minute conditioning time, and an 85 kPa pressure. The instrument parameters are detailed in Table 3. Shimadzu GC solutions software was used for the data processing.

Table 3. Instrumentation Setup: GC-FID parameters for the dual-column detector system.

Parameter	Value	
Inlet temperature (°C)	110	
Injection mode	Split	
Pressure (Kpa)	85	
Column flow (mL/min)	2.78	
Linear velocity (cm/s)	42.30	
Purge flow (mL/min)	3.00	
Split ratio	5.00	
Oven temperature (°C)	40, isothermal	
Detector temperature (°C)	280	
Analysis time (min)	4	
Headspace oven temperature (°C)	60	
Syringe temperature (°C)	70	
Fill volume (mL)	1.75	
Sample volume (mL)	1.00	
Incubation time (min)	5.00	
Sample speed (mL/min)	5.0	
Shaker time (min)	0.50 on	0.10 off
Injection speed (mL/min)	80	

4.3. *Sample Preparation*

An internal standard (ISTD) solution was prepared by adding 25 µL of tertiary butanol to 500 mL of distilled water, with 2.5 g of sodium metabisulphite (antioxidant). Calibrators, quality controls (QCs), and samples were prepared in duplicate by adding 100 µL of sample to 1 mL of internal standard solution in labelled 20 mL headspace vials. Pipetting was carried out using Gilson Pipetman G 20–200 µL and 2–20 µL adjustable-volume pipettes, a Gilson Microman E 10–100 µL positive displacement pipette, and a VWR 100–1000 µL adjustable-volume pipette.

4.4. *Headspace GC-FID Method*

The prepared samples were conditioned according to the headspace sampling parameters (Table 3). The optimised separation conditions consisted of 40 °C GC oven temperature, a 5-min equilibration time with shaking, and an 85 kPa pressure. The syringe was heated to 70 °C, and a 1 mL sample was injected onto the columns. These parameters allowed for a sub-five-minute sample run time with a full resolution and separation of all compounds.

5. Conclusions

The method meets the requirements for our analysis—that is, the analysis of "B" samples. In the UK, a single specimen must be taken. That specimen is split into two, and drivers may request to be provided with one of the parts—the B sample—for independent analysis. Although no direct comparison was made with other headspace sampling techniques, such as loop-filling and pressure-balance systems, this gas-tight syringe method meets the requirements for the alcohol analysis of road traffic blood samples taken in England and Wales, and meets the specifications set in the California Code of Regulations. Gas-tight syringe headspace sampling is a viable alternative sampling technique.

Supplementary Materials: The following supporting information can be downloaded at: https://www.mdpi.com/article/10.3390/molecules27154771/s1, Figure S1: Representative chromatograms of (A) Ethanol negative control samples with tertiary butanol internal standard, (B) QC samples spiked with 20 mg/100 mL of ethanol and (C) a casework samples detected at 79 mg/100 mL.

Author Contributions: L.T. conducted the experimental analysis and manuscript composition. L.S. carried out the experimental analysis. T.B. and V.R. conducted the GC-FID alcohol analysis. H.J.M. performed phlebotomy procedures and experimental design. J.B. and B.K.T. were responsible for manuscript composition, while J.F. conducted the statistical analysis. G.T. and B.R. designed the study and composed the manuscript. All authors have read and agreed to the published version of the manuscript.

Funding: This research received no external funding.

Institutional Review Board Statement: The study was conducted in accordance with the Declaration of Helsinki, and approved by the Institutional Review Board (or Ethics Committee) of Kingston University London (protocol code 1819063.1 on October 2018).

Informed Consent Statement: Informed consent was obtained from all subjects involved in the study.

Conflicts of Interest: The authors declare no conflict of interest.

Sample Availability: Samples are not available from the authors.

References

1. Felson, R.; Staff, J. The effects of alcohol intoxication on violent versus other offending. *Crim. Justice Behav.* **2012**, *37*, 1343–1360. [CrossRef]
2. Scott-Ham, M.; Burton, F.C. Toxicological findings in cases of alleged drug-facilitated sexual assault in the United Kingdom over a 3-year period. *J. Clin. Forensic Med.* **2005**, *12*, 175–186. [CrossRef] [PubMed]
3. Dubowski, K. *Manual for Analysis of Ethanol in Biological Liquids*; Department of Transport: Washington, DC, USA, 1977.
4. Legislation.gov.uk. The Road Traffic Act 1988 (Prescribed Limit) (Scotland) Regulations. 2014. Available online: https://www.legislation.gov.uk/sdsi/2014/9780111024478 (accessed on 18 February 2022).
5. Legislation.gov.uk. Road Traffic (Amendment) Act (Northern Ireland). 2016. Available online: https://www.legislation.gov.uk/nia/2016/11 (accessed on 18 February 2022).
6. Legislation.gov.uk. Road Traffic Act 1988. 2011. Available online: https://www.legislation.gov.uk/ukpga/1988/52/section/5 (accessed on 18 February 2022).
7. North, P. *Report of the Review of Drink and Drug Driving Law*; DfT Publications: London, UK, 2010.
8. Logan, B.K. Analysis of alcohol and other volatiles. In *Gas Chromatography in Forensic Science*; Tebbett, I., Ed.; Ellis Horwood: New York, NY, USA, 1992; pp. 87–108.
9. Kellner, R.; Mermet, J.-M.; Otto, M.; Varcárcel, M.; Widmer, H.M. *Analytical Chemistry: A Modern Approach to Analytical Science*, 2nd ed.; Wiley-VCH: Weinheim, Germany, 2004.
10. Kovatsi, L.; Giannakis, D.; Arzoglou, V.; Samanidou, V. Development and validation of a direct headspace GC-FID method for the determination of sevoflurane, desflurane and other volatile compounds of forensic interest in biological fluids: Application on clinical and post-mortem samples. *J. Sep. Sci.* **2011**, *34*, 1004–1010. [CrossRef] [PubMed]
11. Jones, A.W.; Holmgren, A.; Ahlner, J. Toxicological analysis of blood and urine samples from female victims of alleged sexual assault. *Clin. Toxicol.* **2012**, *50*, 555–561. [CrossRef] [PubMed]
12. Brown, D.J.; Long, W.C. Quality Control in Blood Alcohol Analysis: Simultaneous Quantitation and Confirmation. *J. Anal. Toxicol.* **1988**, *12*, 279–283. [CrossRef] [PubMed]
13. Zuba, D.; Parczewski, A.; Reichenbacher, M. Optimisation of solid-phase microextraction conditions for gas chromatographic determination of ethanol and other volatile compounds in blood. *J. Chromatogr. B* **2002**, *773*, 75–82. [CrossRef]
14. De Martinis, B.S.; Martin, C.C.S. Automated headspace solid-phase microextraction and capillary gas chromatography analysis of ethanol in postmortem specimens. *Forensic Sci. Int.* **2002**, *182*, 115–119. [CrossRef]
15. De Martinis, B.S.; Martins Ruzzene, M.A.; Santos Martin, C.C. Determination of ethanol in human blood and urine by automated headspace solid-phase microextraction and capillary gas chromatography. *Anal. Chim. Acta* **2004**, *522*, 163–168. [CrossRef]
16. Westland, J.L.; Dorman, F. Comparison of SPME and static headspace analysis of blood alcohol concentration utilizing two novel chromatographic stationary phases. *Forensic Sci. Int.* **2013**, *231*, e50–e56. [CrossRef] [PubMed]
17. California Code of Regulations. Public Health (17) § 1220.1. 2021. Available online: https://govt.westlaw.com/calregs/Browse/Home/California/CaliforniaCodeofRegulations?guid=I8308083A5BD746EC87F10B98F14121AA&transitionType=Default&contextData=%28sc.Default%29 (accessed on 18 February 2022).

18. Taylor, L.J.; Remeskevcius, V.; Saskoy, L.; Brodie, T.; Mahmud, J.; Moir, H.; Brouner, J.; Howe, C.; Thatti, B.; O'Connell, S.; et al. Determination of ethanol in micro-volumes of blood by headspace gas chromatography: Statistical comparison between capillary and venous sampling sites. *Med. Sci. Law* **2021**, *61*, 86–96, ISSN: 0025-8024. (*in print*). [CrossRef] [PubMed]
19. Elliott, S.P.; Stephen, D.W.S.; Paterson, S. The United Kingdom and Ireland association of forensic toxicologists: Forensic toxicology laboratory guidelines. *Sci. Justice* **2018**, *58*, 335–345. [CrossRef] [PubMed]
20. Mihretu, L.D.; Gebru, A.G.; Mekonnen, K.N.; Asgedom, A.G.; Desta, Y.H. Determination of ethanol in blood using headspace gas chromatography with flameionization detector (HS-GC-FID): Validation of a method. *Cogent Chem.* **2020**, *6*, 1760187. [CrossRef]
21. Monteiro, C.; Proença, P.; Tavares, C.; Castañera, A.; Corte Real, F. Interference of anesthetics in blood alcohol analysis by HS-GC-FID: A case report. *Forensic Sci. Int.* **2016**, *265*, 65–69. [CrossRef] [PubMed]
22. Neelamegam, R.; Ezhilan, B. GC-MS analysis of phytocomponents in the ethanol extract of *Polygonum chinense* L. *Pharmacogn. Res.* **2012**, *4*, 11. [CrossRef] [PubMed]
23. Magistrates' Court Sentencing Guidelines. 2021. Available online: https://www.sentencingcouncil.org.uk/the-magistratescourt-sentencing-guidelines (accessed on 18 February 2022).

MDPI
St. Alban-Anlage 66
4052 Basel
Switzerland
www.mdpi.com

Molecules Editorial Office
E-mail: molecules@mdpi.com
www.mdpi.com/journal/molecules